ACPL ITEM
DISCARDED

3 1833 01553 8780

333.75 M13L
Manning, Richard, 1951-
Last stand

D1442386

DO NOT REMOVE
CARDS FROM POCKET

12/13/91

ALLEN COUNTY PUBLIC LIBRARY

FORT WAYNE, INDIANA 46802

You may return this book to any agency, branch,
or bookmobile of the Allen County Public Library.

DEMCO

Last Stand

Last Stand

Logging, Journalism, and the Case for Humility

Richard Manning

PEREGRINE SMITH BOOKS

SALT LAKE CITY

Allen County Public Library
Ft. Wayne, Indiana

First edition

Copyright © 1991 by Richard Manning

96 95 94 93 92   10 9 8 7 6 5 4 3 2 1

All rights reserved. No part of this book may be reproduced in any manner whatsoever without written permission from the publisher.

This is a Peregrine Smith Book, published by Gibbs Smith, Publisher, P.O. Box 667, Layton, Utah 84041

Design by Kathleen Timmerman
Cover photographs by Michael Gallacher

Manufactured in the United States of America

Printed on recycled paper

Library of Congress Cataloging-in-Publication Data

Manning, Richard, 1951-
    Last stand : logging, journalism, and the case for humility /
Richard Manning.
        p.   cm.
    Includes bibliographical references.
    ISBN 0-87905-389-5
    1. Logging—Northwest, Pacific. 2. Deforestation—Northwest,
Pacific. 3. Manning, Richard, 1951–   . 4. Forest products
industry—Northwest, Pacific. 5. Forests and forestry—Northwest,
Pacific. 6. United States. Forest Service. 7. Journalists—United
States—Biography. I. Title.
SD538.2.N75M36   1991
333.75'137'0973—dc20   91-3505
                                                          CIP

DEDICATION

This is for my son, Josh.
This is for our unwarranted
but necessary
faith in the future.

EPIGRAPH

Witness

I want to tell what the forests
were like

I will have to speak
in a forgotten language

W.S. Merwin

ACKNOWLEDGMENTS

WHILE THE BLAME for the execution of this work is simply mine, credit for the idea goes to William Kittredge. His encouragement, advice and assistance were crucial. I am in his debt, a mortgage I share with a whole bevy of Western writers. Where would we be without Bill? In particular, I will always rely on a bit of uniquely Kittredgian instruction that settled this reporter's fears on tackling my first book: "Write it just like one of your newspaper stories only make it longer," he said.

The book's path also crossed at crucial moments the paths of two other writers. Richard Nelson came along just as ideas were beginning to jell into words and provided an important push to the process. Then after some good ideas had jelled into some bad words — mine — Peter Matthiessen delivered a stern hiding that produced at once the humility and confidence this work so needed. He is as fine a teacher as he is a writer, a combination mostly responsible for any quality this book may have.

In the same vein, I wish to thank Henry Harrington and the Teller Wildlife Refuge for backing an enterprise so thoroughly seditious as an environmental writing institute. It was the institute that put me in contact with Matthiessen.

Before there was writing, there was reporting. My city editor, Brian Howell, was a hero, performing the delicate task of keeping me employed while steering a charged piece of reporting through the strange bureaucracy of the *Missoulian*. Both Michael Gallacher and Kurt Wilson were as solid as partners/photographers/eyes as a reporter could ask for. I thank them for their insights through the years. Especially, I thank Michael for his weird mix of insanity, energy and friendship that sustain me to this day. And most particularly, I value the support, criticism, and love that came from my colleagues at the *Missoulian* through the fiery days of the reporting, through the exercise of the craft we all value, and through my split from the family, painful as it was. Michael Moore, Kathleen Kimble, Jim Ludwick, Mea Andrews, Michael McInally, and Pat Sullivan are people of integrity, journalists, and friends.

My reporting also was greatly aided by the staff of the Lolo National Forest, a progressive bunch in a sometimes backward outfit. I am particularly indebted to Vic Applegate, Mike Hillis, Chuck Spoon, Orville Daniels, and to my friend Marcia Hogan, who I still think knew where this was headed before I did. Thanks also to Arnie Bolle for his technical advice and his serving as an example to all of us who value the forests. Barry Dutton's expertise on soils was likewise invaluable.

On the other end of the project, there were important contributors to getting this into print. John Mitchell gave me both professional support and practical help. Among other boosts, John pointed me to Tim Schaffner, the agent I needed. Tim believed enough in the environment to take a chance on a quirky project like this. Gibbs Smith believed, too, with enough conviction to risk his perfectly good money on an unproven writer. Both I and my landlord thank him for this. All of us are indebted to Gibbs for his commitment to the environment and his defiance of the corporations through the years.

Heather Bennett was my editor and managed to find a lot more rough edges than I remember installing. Much of the credit for any quality in this work goes to her. Also saving me from some egregious errors were Marcia Hogan and Woody Kipp. I thank them for their advice on the manuscript.

Lingering behind the writing of this book was the living of it. Therein lurks an enormous debt that I can never repay. The shaping of some of the deeper notions that drive my thinking occurred during the course of that part of my life, half of my life, shared with one of the world's finest humans. Convention suggests that I not unearth this matter now, but honesty suggests otherwise. Thank you, Margaret, for the wisdom of your heart and mind.

Finally, I wish to make something of the same point that the book does: acknowledge the value of the forces that sustain us through each day. My life these days has found the joy of the light in the forest largely through the love of my wife, Tracy. She taught that even old guys deserve some happiness. Her warmth and optimism are generously layered between every line.

CHAPTER ONE

IN NORTHERN MICHIGAN there is a swamp where my grandfather cut trees with an ax. One day my brother and I found him there, in a clearing his work had opened to sunlight. He spotted us immediately, because his bright eyes missed nothing in his forest. Our interruption was an excuse for him to stop and hone his ax on a whetstone he kept in the chest pocket of his buffalo-plaid shirt. He spat at the center of the puck-like stone that years had rubbed baby-ass smooth, then he honed. The woods rang with the conviction of steel on rock. Then the whack of the ax beat on, and my grandfather, a quiet man, cut trees. My brother and I watched, cheering the chips' flights through the swamp's slanted light. His skill was magic to a couple of grade-school kids. We had already tried this trick of chopping. We had managed to get our own hands on an ax, or more precisely, an army-surplus hatchet we had hidden from our mother. Our chopping was not nearly so impressive, though. We bludgeoned away whole afternoons, never managing to topple more than a sapling.

Our own grandfather, though, could make great chunks fly, and so we believed our clumsiness a temporary handicap of childhood. We believed ourselves kin of Paul Bunyan, likely lost but legitimate heirs to an entire herd of blue oxen. The Paul Bunyan stories still circulated in Michigan's northern Lower Peninsula then, which was a wry irony. It was the early sixties and the big trees were all gone. The state had been shaved like a dog, surrendering the land to dairy farms, potato fields, and gravel pits. The vestigial forests were second- and third-growth deciduous trees mixed with spruce, cedar, and tamarack swamps. They were sticks, poor cousins of the pines that had spired that place since the glaciers pooled the Great Lakes. The trees were gone, but the Paul Bunyan myth survived for a reason: It tells our story. My people, our people, are loggers, a people of trees. Our nation was founded on and is now foundering on this ancient enterprise. I share the guilt of this logging. My complicity is as demonstrable as my grandfather's skill with an ax, as my writing this on paper, as my love of needle-grained Sitka spruce on the face of an old guitar.

When my brother and I first learned of our past, my grandfather was cutting cedar, a sign of reduced circumstances for us and for the once regal forests of Michigan. "Logging" in Michigan was and is a misnomer; there are no more logs. Now there is pulp: adolescent popple, birch, and other fast-growing deciduous trees that are pulverized to paper and fake wood-chip paneling.

When there were actual logs, a century ago, they came from the white pine that once made Michigan, Minnesota, and Wisconsin the heart of the nation's lumbering, which was itself the largest industry in the nineteenth century. In this latter half of the twentieth, men like my grandfather were reduced to picking up a few dollars cutting and peeling eight-foot cedar fence posts and selling the evergreen boughs to florists.

In northern Michigan, bough cutters were called "cedar savages," mostly because they were poor. We were not poor, I guess, but close enough. I didn't mind. I considered it a privilege to work in the woods. It was my grandfather's own woods, an eighty-acre plot logged at least once, but still holding its second growth. My grandfather and his ax had been at work for years, but it never seemed to change the way the woods looked. Cutting trees was just something he did, my mother's father.

His name was George Mayo, an Anglicized version of the original French Maillioux. He had held other jobs. He had trapped beaver, mink, and rabbits in the state's Upper Peninsula. He had farmed. He had worked in the "shops down below," northern Michigan's term for the auto factories clustered around Flint and Detroit. Some piece of machinery there bent one of his fingers sideways at forty-five degrees, leaving him with a bum hand that was a source of constant fascination for the grandkids.

In his woods, he seemed to fit like a moss-layered stump. He poached deer, even fawns, in the summer because they were the "best eating." He snared rabbits in illegal copper wire loops and stalked morel mushrooms in the spring. He caught pike with a long cane pole, smoked cigarettes over the protests of my Baptist grandmother, and in the afternoons watched soap operas on television and cried. He came from a time and place where an orange was considered a good Christmas present. When I went to college to study political science, he explained that the one fact I needed to know about politics is that all Republicans are Communists.

He was, however, not at all confused about the theory and practice of the woods. His place was a swamp, choked with the brush of a young deciduous forest. When I was a kid, my body was still too big and clumsy to thread his forest, and I had to tunnel like a rabbit next to the ground where the limbs couldn't grow. My grandfather, though, would slide through the undergrowth like an old trout slips through his home creek. He was short, quick, and half Chippewa.

He cut trees, but he was still a part of the woods. He lived there like a creature the trees protected. This makes me unusual, I guess, simply because most of us cannot claim ancestors who liked the forest. Some of our people lived there, but most despised or feared the forest. My other grandfather, my father's father, was squarely in this mainstream. He hated trees.

Alexis de Tocqueville observed, "Europeans think a lot about the wild open spaces of America, but the Americans themselves hardly give it a thought. The wonders of inanimate nature leave them cold, and one may almost say they did not see the marvelous forests surrounding them until they began to fall beneath the ax." Here is a deep streak in our national character. Forests were first viewed as an impediment to farming every bit of soil that would hold a seed. Only later were they considered a commodity to be converted to fuel, ships, and houses.

More deeply, though, we believed the forests to be the savagery that held wildness. They were the netherworld so despised by our puritanical streak. They were suspect to the Christian mind as pagan, Druidic, and in American terms, native. The Puritan Michael Wigglesworth in 1662 called New England's forests "a waste & howling wilderness where none inhabited but hellish fiends & brutish men."

This was the majority view. Two centuries after Wigglesworth lived, though, a minority opinion began to germinate with the rise of the romantics and the transcendentalists. Still, this was a glorification of a nature that had been tamed a bit. People like Henry David Thoreau often venerated the pastoral, not the wild. When he left the relative safety of the family pond and encountered his first bit of real wilderness, the untouched forests of Maine, Thoreau pronounced them "savage and dreary."

Then, as now, singing the wonders of nature was largely the work of a leisure class. The working people who faced the raw edge of the American frontier held with the Puritans, with the doctrine of salvation at the head of an ax.

MY PEOPLE ON MY FATHER'S SIDE WERE "STATE OF MAINERS," a fact that wound my paternal lineage tightly to the lumber trade and, I suspect, made my Grandfather Manning hate trees. The northern tier of America's deforestation fought its first key skirmish in Maine, where waves of Scandinavian, French, German, and in my case, English immigrants attacked the green massif of white pine. Pine was the tree of choice then, which traces to its designation as the "king's tree." The continent's first forest reserves were grounded not in conservation but in warfare. Because the straight, tall white pine made the best ships' masts and planks, naval primacy was a matter of access to looming forests. Before the revolution, these trees were reserved for the king of England and His Majesty's Royal Navy. The nation's federal forests were labeled naval reserves until the middle of the last century.

Shipbuilding and, later, construction of the railroads and of the cities they fostered built a voracious lumber industry in Maine and the rest of the Northeast. As with other commodities, though, there was no thought of running out of white pine. There were always more trees to the west a ways. Even before the Civil War, the nation's lumber industry began to migrate from the once seemingly endless stretches of white pine in Maine to the then seemingly endless stretches of white pine in the upper Midwest.

Caroline Kirkland, an educated New Yorker and a pioneer in Michigan, said:

> The Western settler looks upon these earth-born columns and verdant roofs and towers which they support as "heavy timber,"—nothing more. He sees in them only obstacles, which must be removed, at whatever sacrifice, to make way for mills, stores, blacksmith's shops—perhaps a church—certainly taverns. "Clearing" is his daily thought and nightly dream; and so literally does he act upon this guiding idea, that not one tree, not so much as a bush, of natural growth must be suffered to cumber the ground, or he fancies his work

incomplete. The very notion of advancement of civilization or prosperity seems inseparably connected with the total extirpation of the forest.

A Maine congressman lamented the loss of his state's best loggers to the virgin forests of Michigan: "Most fortunate that state or territory which shall receive the largest accession of them, for like the renowned men of olden times, 'they are famous for lifting up the axes upon thick trees.'"

When I was a kid growing up in Michigan, I saw old white pine. An aging philanthropist realized he owned an unlogged plot of stately trees near the shore of Lake Huron, so he gave the land to the state. I was in the high school band that suited up for the occasion of the state's acknowledging his gift. Organizers of the ceremony bused us the thirty or so miles north of town to render a few Sousa marches to the only big, old trees most of us had ever seen. In the 1840s, Michigan had been covered with ancient white pine like wheat covers the Dakotas. In the 1960s, when someone found a relict grove, we fielded a brass band.

There is precedent for this behavior. As classical forests grew scarce, the Romans set aside sacred preserves north of town and used to haul people in for various ceremonies that featured the age's counterpart to Sousa marches. Even earlier, Plato's Laws spelled out penalties for violating such sacred groves. In the era of Christian fanaticism that followed the Roman Empire, however, loggers prevailed. Preservation of groves of sacred trees was considered pagan pantheism. Deforestation revived the veneration of trees in the Christian Midwest, although it is not clear whether Sousa marches constitute pagan worship of the white pine. Then, though, I thought them to be grand trees, worthy of my saxophone's squawking homage.

Before then, I had seen a big white pine in a picture. My Grandfather Manning looked oddly young in the photo I am thinking of, probably still in his twenties, still had both of his eyes. He was smiling, something he almost never did. He and another guy had cut a tree. He was not, however, a woodsman; he was a farmer.

Once loggers had stripped the trees from the rolling, glacial till smeared on northern Michigan, some people staked out the remaining sand pits and sour soil moraines and called them farms. Alpena County, where I was raised, is one great and uniformly flat boggy plain broken only by a large glacial deposit on its western edge. This esker came to be known as Manning Hill, largely because my great-great-grandfather beat that gravel hill into something of a farm.

My grandfather continued the war with the soil, waging it with a stubbornness that was local legend. A large subset of that legend was his total disdain for trees. He would tolerate on his three-hundred-acre domain only one small woodlot large enough to feed the basement furnace of the insulbrick-sided farmhouse. He is alleged to have once mowed down all of the farm's orchards simply because they too much resembled a forest. When anyone mentioned pruning, he would stipulate that all trees were best pruned to leave a short stump. Then he'd grin. The legend says this attitude rooted during my grandfather's youth when that same farm was mostly orchard. It was his job to cultivate the trees with horse-drawn equipment, but when the horses brushed against them, sap and apples tarnished the harness's hames. In my grandfather's leisure hours, he polished the hames. This confirmed his hatred of trees. My grandmother was allowed to plant nothing more looming than a hollyhock in her yard. Grandfather Manning's land was for sweat, potatoes, oats, and hay, and none of his four sons could tolerate it. All took city jobs as soon as they were able.

My father left for a string of jobs, but then he never really left the farm's attitudes. Neither have I. We are not unique in this; our lives are evidence of the way people lived in the

Midwest at the time: with equal parts hubris, sweat, and aching fear learned from the Great Depression. In practical terms, these pressures meant you by God learned to work. Our essential relationships were with the soil or the woods or with the gravel and rocks we used to build houses, but all of these relationships we prosecuted with work. We were raised to understand early that we were on earth to sweat, lift and build and very little else. Personal worth was solely weighed on a scale whose high side was bounded by the judgment, "He is a good worker." This was the best that could be said of someone in the world staked out by my Grandfather Manning's glower and hard hand.

Most people feared that glare steered by his one good eye. He had lost the other eye in a fall from a barn roof many years before I was born. By the time I knew him, his socket muscles had deteriorated, and they no longer held a glass eye. He wore glasses with one side frosted and shielded to cover the hole. This and his curmudgeon's ways scared most kids, including me until the time I was about ten years old. Then my father started taking me to visit my grandfather on report-card days. Wordless, he would check to see that I had earned straight A's, then pry a dollar bill from a coinpurse in his pocket and give it to me. Likely it was the first time he had given a kid money for anything other than stoop labor. It made me proud. I could earn his approval with something other than calluses. Nobody admitted to minding his work then, because that was what life offered. Still, no one seemed to want his kids to have the same life, and in a simple world, education was viewed as the escape. There was a lot of talk about education when I was growing up. It was assumed I would not live as my father and grandfathers had lived.

A couple of years before my Grandfather Manning died, my Dad ended a series of jobs that had moved us away from Michigan. He decided to move back and build a house on sixty acres my grandfather had sliced off from the farm. It was too small to work as a farm, nor did my father seem to have much inclination to do so. There were other jobs, construction, concrete work, truck driving, and so on. Some years it went all right. When it didn't, we often wound up in the woods, my other grandfather's woods, a parcel of eighty acres that lay just two miles southeast of Manning Hill. There were still a few dollars to be made on cedar and pulp when times were lean, and so that's what we did. A few years before I entered high school, we bought a creaky little Oliver bulldozer that was older than I was, and after school, it was my job to run the cat. My brother and I would load a sledge with cedar poles, then skid them to a landing where the trucks could get them. It wasn't bad work.

During the next few years, my Dad got more involved in the woods, buying bigger cats and finally a small portable sawmill that would feed on dying elm. Much of Michigan's old-growth white pine had been replaced by American elm that by the 1960s was beginning to stand stately. Then the imported Dutch elm disease hit the state, killing great swaths of forest. A dead tree is and was considered a waste, and so the elms were being cut as rapidly as possible. My Dad's sawmill helped in that project. It was a short-lived enterprise, though, because one of those trees fell across the leg of a guy who was running a skidder. My father was never much for handling details such as buying worker's compensation insurance. There was some legal nastiness, and he went broke, off to a new business.

That was about the time I headed in a different direction. I was doing all right in school. My grades, combined with the late-sixties liberalism that created scholarships for kids who needed them, earned me a slot at the University of Michigan. I enrolled in 1969, intending to become a doctor, but those were the anti-war years. Ann Arbor was alive with rebellion, and I was nothing if not a rebel. Swept along on this wave of rage, I went first into the study of politics, the more radical the better, then finally settled to a career in journalism. The job eventually would bring me deep into the Rocky Mountain West.

The first wave of Midwesterners to come farther West, to leave the dwindling white pine forests and cross the plains into the pine and fir forests of the Northwest, did so on the Oregon Trail. There was a beacon on that trail, a single tree that marked the western edge of what was then called the Great American Desert. Plains-weary, woods-raised settlers would spot this tree and point their wagons on it like ships on a light. This lone welcoming tree told them a desert had been crossed and they were headed for something that looked like home. The Oregon Trail had existed only a few years when the explorer John C. Fremont went looking for that tree. He found it had been dropped by an immigrant's ax, left at the fringe of a now-more-treeless plain. Such is our conflict with the trees that guide us, evident in the fate of Fremont's tree and evident too in the way a transplanted Midwesterner sees the scrawny Rocky Mountain maple.

The big trees of the West are almost exclusively conifers: ponderosa pine, Western larch, and Douglas fir. A few deciduous trees—cottonwoods and quaking aspens—hug the creek sides and canal banks, but these are not forests. If you are hit by the tall light filtered by great trees, then you stand among conifers. I imagine this is what Maine or Michigan must have felt like during the days of the white pine, when the first green boughs stood a good sixty feet overhead and the forests were pillared with spars. Still, this Western forest was new to me. Like most Midwesterners, I had been raised among maples and oaks. Not spars but candelabras were home and peace. As much as I love the West, I miss the crunch of fall's leaves.

Along the creek beds of the Rockies, there are maples of sorts, mountain maples, the same fingered leaf that the Midwest produces. The same flush of color at year's dying. These

Western maples, though, are not really trees but shrubs, wispy memories of Midwestern forests. Still they strike a chord. When I see these shrubs, I wonder how much of our lives is encoded in trees, how my predecessor immigrants to the Northwest must have first felt about the mountain maple. They did not come here for that shrub; those early migrants came first for the white pine that grew in western Montana and North Idaho. First Maine, then Michigan and now here, they knew the white pine best. My father's mother was a Richardson, one of the few of the old lumbering families left in northern Michigan. She was stranded in the Midwest by her marriage to a farmer. When I was a kid, my father told me her people had gone West where the big pine grew.

When I came west in 1979, I figured I was following them, just a late stage in the migration pulled along by trees. I know that my people came before me, because some are the sort that would have cut Fremont's tree. They would have found it beautiful and beckoning across that great plain, maybe for a moment even mystical, but ultimately simply useful, most useful dead.

I know my people came before me because when I leave the brushy creek beds and those shrub maples of the West, I come to towns such as mine. The streets are lined with large Midwestern-type maples, planted trees, illusions of the permanence of forests. The towns of the West look like the towns of the Midwest: tree-lined streets, clipped lawns, and kids diving into the fall's sheddings of maples. Clapboard houses, picket fences, Presbyterian churches, all civilizing the place of buffalo, Blackfeet, and sage. Maples civilizing pine.

I know my people were here before me because they are the sort who would have moved West to the call of good jobs and tall trees. They would have run the crosscut saws, bandsaws, big-wheel skidders and planers. They are the sort who would have spent a day cutting pine and then, after spotting a mountain maple at the tail of a draw, would have become homesick for the trees they knew. They would have sent east for maples and planted them in neat rows along the front stoop.

IN 1833, CARPENTERS WORKING ON A CHURCH IN CHICAGO developed a technique called balloon framing, accelerating America's timber age. This was the beginning of the studs and joists that frame virtually all American homes today. Before then, houses were made of logs or stone or great timbers sawed and meticulously joined on site. Balloon framing meant relatively small amounts of light pieces of lumber could be transported great distances so that relatively unskilled carpenters could build houses. Colonizing the treeless plains became a possibility. The railroads with their wood-fueled engines and roads laddered with wood cross ties began hauling settlers and sawed lumber into the Plains. Per capita consumption of wood leaped four hundred percent in the first half of the nineteenth century. It would quadruple again in the second half.

I imagine this time as a movie starring people I knew from childhood, people who cut trees and built things. I conjure images flushed with activity: tree felling, sawing, and building, all with an exuberance that did not question our right to do so. I try to understand how all of this era shaped me, derived as I was from my grandfathers and their axes. There is conflict; today I see limits. There is an ocean just beyond the last of the big trees. There is no more West, and so sweat, labor, and building are no longer pure pleasures.

Oceans notwithstanding, though, our generation's lust for construction pales our grandfathers'. We think of our ancestors as the cutters and sawyers and builders, yet their generations and all before came not even close to our time's ability to build to every horizon. The acceleration that balloon framing began in the early 1800s has not slowed. In 1869, when timber led the national economy, when our nation sailed, railed, heated, and built near everything with wood, when the rivers of Maine and the upper Midwest were awash with logs and loggers, when the live-oak and yellow-pine forests of the South were still being felled pell mell for building sailing ships, the entire nation produced about thirteen billion board feet of lumber annually. By 1990, the four states of the Pacific Northwest by themselves produced more than twenty billion board feet of softwood lumber a year, about sixty-five percent more than the whole nation produced at the height of the timber age. Per capita consumption and the total cut peaked in about 1910, but the zenith was to a large degree driven by fuelwood, which we have mostly eliminated. This is the age of sprawling suburban homes, second homes, a society of throw-away paper packaging. Per capita consumption has fallen in this century, but increased population and habits have compensated. In the Northwest, this is still the timber age.

The mythology of our time as it is presented on our postcards, coffee-table books, tree-lined streets, and public relations campaigns is that we are a nation that loves trees. The reality is that we are a nation that cuts its trees. This is not a habit that began with our cedar-sided vacation homes or even that began with our nation.

CHINA WAS EFFECTIVELY DEFORESTED MORE THAN A THOUSAND years ago. India followed a couple of hundred years later. The line of deforestation coupled with a line of thought that eventually marched into our nation, however, did not begin in India and China. It began in the Middle East, in the east with a line headed west. We can trace this line as the spread of civilization, culture, and religion, or we can trace it as the advance of deforestation and desertification.

Judaism, Christianity, and Islam all rose from what were once forested lands. Turkey was covered with oak. Syria was renowned for its forests. Lebanon was a country not of desert, but of cedars that once drove the navies of Phoenecia, Persia and Macedonia. King Solomon was noted for his ability to discuss the merits of the various species of trees.

Ceramic mosaics from the fourth century A.D. in Algeria show lions, panthers, ostriches, hartebeast, oryxes and wild assess, forest beasts once as common as the stands of Algerian cedar and oak that held them. The trees were cut. Since Roman times, Morocco has lost about 12.5 million acres of forests to logging. North Africa was once known as "the land of continuous shade."

As early as 1100 B.C., a pattern began that saw city-states wither after they had denuded their homeland of ship timbers. They would fall to neighbors who could still raise a navy. First fell the Minoans, then Crete. By the fourth century B.C. most of Greece was bare, sending that maritime culture in search of timbers in Italy and even as far west as Spain. Writing twenty-five hundred years ago, Plato speaks of this in words that haunt today:

> Contemporary Attica may accurately be described as a mere relic of the original country. There has been a constant movement of soil away from the high ground and what remains is like a skeleton of a body emaciated by disease. All the rich soil has melted away, leaving a country of skin and bone. Originally the mountains of Attica were heavily forested. Fine trees produced timber suitable for roofing the largest buildings; the roofs hewn from this timber are still in existence [700 years later]. The country produced boundless feed for cattle, there are some mountains that had trees not so very long ago, that now have nothing but bee pastures. The annual rainfall was not lost as it is now through being allowed to run over a denuded surface into the sea, it was absorbed by the ground and stored. . . . The drainage from the high

ground was collected in this way and discharged into the hollows as springs and rivers with abundant flow and a wide territorial distribution. Shrines remain at the sources of dried-up water sources as witness to this.

First examination suggests that deforestation necessarily follows civilization, yet the record argues for a more interesting relationship: deforestation, it would appear, is a necessary precursor of civilization. The argument is this: the forests must fade before man, his plows, and his cities can move in. Plato's leisure for thought, one measure of civilization, was paid for by the death of native forests more than a half century before his time.

The Greek sophist Secundus, in a line that my farming grandfather would have endorsed, writes of "agriculture on the march with the torch and the ax." Cities are built with trees and, just as importantly, are fed only in the absence of forests. At the same time, cities still need those forests they have depleted. Civilization seems first to cut its native forests and then reach abroad for others until it can reach no further. It was the power of the navies that drove this equation from well before Plato's time almost until ours. The Greeks eventually reached to southern Italy, and Spain "thick with woods and gigantic trees," until Greece could reach no more. Then Rome reached. The Greeks already had denuded the south half of Italy, so the Romans logged Spain and North Africa of naval timber, fuel, and decorative hardwoods. Then Rome collapsed.

Spain eventually rose to become a naval power, an empire, but by the time it did so, in the fifteenth century, it was already deforested by its predecessor empires. Spain imported ship timbers from Scandinavia. The French navy during the same period and into the nineteenth century pulled logs from northern Italy and Albania. Native forests of Mediterranean

France were long since gone, sacrificed to city building that allowed France to become a naval power in the first place. The British Isles were once lush with oak and beech, but by the close of its wooden era, half the logs in British ship yards came from northern Italy. The foreboding forests of Europe had sheltered man since our genus first hit the continent, from the days of Neanderthal until Medieval times, several hundred thousand years. The forests produced the frame of mythology that survives in Western culture today. Now they are reduced to a few dying tree farms in Germany.

Driven by the lust to build boats, the line moved west across the land. I am a part of this line. The great forests of Europe were gone by the time my people left England for Maine, and now the white pine of Maine are gone, as they are in the upper Midwest. Now I stand here in the Northwest among the pine and fir that are falling all around me. Mostly I see them stacked on trucks headed for the mills that punctuate every drainage in the Pacific Northwest: Washington, Oregon, North Idaho, western Montana, British Columbia, northern California, and Alaska. Now the line cannot move west, hemmed by the Pacific Ocean, and beyond, hemmed by the deforested deserts of Asia, where the soil has been so damaged that the Chinese fir no longer will grow. There our linear logic becomes circular. The line of forests does not stretch infinitely off into space but arcs straight back to face our beginnings. At the beginnings of this great circle, we meet ourselves.

My grandfathers are dead now, disappeared. Their marks on me are hidden deep in psyche and gene, so I can hide and deny my legacy of logging, my complicity. Most of us can, standing as we do a generation or two or a state or two removed from the people who sweat and saw. Living as we do among white plastered walls that hide the web of lumber studs within. Our houses lie hidden behind a band of maple trees, imported and planted to conjure the peace of a gone forest. Now, though, as we catch clear sight of the ocean to our west, it appears that this conjuring of the ghost forests will soon be all that remains. Soon we will be reduced to accepting a row of planted trees as substitutes for the web of mystery we call a forest. We struck this bargain in pursuit of wealth, but if we follow through, it shall become the ultimate measure of our poverty. Given this prospect, it seems time to cease denying our legacy and future, to face the consequences of what we and our grandparents have done.

CHAPTER TWO

MY CENTER OF GRAVITY in the world's circle of deforestation is a series of newspaper stories I wrote during the spring of 1988. I think of this center, though, not so much as a point but as a peephole in a board fence: the closer my eye approaches it, the wider becomes the view of what lies beyond. I was a confident reporter when I first put my eye to the hole, confident I knew where this looking would lead. In the end, though, the field widened far beyond my expectations. It shattered a career and settled a life. Now, in this retelling of my story, this widening must proceed on a parallel path, a design set by the forest. I begin at the narrow center with a simple act of reporting.

This story came to me during the first months of 1988. I took a new assignment as the environmental reporter for the *Missoulian,* which is the daily paper based in Missoula. One of the three largest papers in the state, it covers all of northwestern Montana, an area the size of West Virginia. I had been with the paper almost three years, reporting until then on state politics and county government. It was my fifth paper in fourteen years as a reporter. Earlier I had worked in

Michigan, then for about five years at three papers in southern Idaho. I had moved to Montana in 1985 largely for a chance to live among the state's mountain streams and roadless backcountry. A few years later, when the chance came to switch beats to write about such places, I took it.

Steve Woodruff, the previous environmental reporter, had become editor of the paper's opinion page. It was really Woodruff who put me onto the project that triggers all of this. Years before, he had done some interviews, but had been unable to turn the story. He gave me a computer printout of one interview and steered me toward one of his anonymous sources. That was the beginning of what would unfold as a major project of investigative reporting.

This source is a bureaucrat and a forester, a person schooled to cut trees. Some people call foresters "timber beasts," because they are considered as predisposed to cutting trees as soldiers are to waging wars. This particular forester has presided over the logging of a large corner of Montana, but in doing so, he came to understand that errors were made. He has allowed his love of the land to open his head to science's growing wisdom about forests. This wisdom suggests that some of his colleagues have harmed the land.

I met him for lunch once, just to become acquainted, but did not ask him then about abusive logging on corporate property. Just two corporations—Plum Creek Timber Company and Champion International—own and log on about 1.7 million acres scattered across western Montana. Largely, this land exists as twenty-six hundred separate square-mile sections; taken together, an area larger than Delaware. It was common knowledge the corporations had logged hard in recent years, but no one outside of the corporations really knew how hard. Corporate lands are flung far from highways and up isolated canyons, unseen. Few people thought there was much to discover there beyond the obvious: the corporations were abusing their land more than most people would prefer, but it was, after all, their land. Some people, however, suspected there was more to the story.

Encouraged by Woodruff, I followed up on that luncheon with a meeting in the source's office. Without much prodding, he laid out the logical framework that pointed to a clear conclusion: the logging business in Montana had taken a brutal turn that would all at once punish the land, the local economy, and the small-time loggers and mills. This was not, however, a case of corporate predation as usual. Instead, some unintended spin-offs of environmental law, a national business climate characterized by hostile corporate takeovers, and some land ownership patterns laid down more than a century ago were forming the jaws of a vise. It hinged on open abandonment of "sustained yield." This is an article of faith that has organized forestry since Gifford Pinchot, the first head of the Forest Service, brought the science to this country from Europe near the turn of the century. The doctrine is simple: it says that one does not cut timber faster than natural growth replaces it. If it takes sixty years to grow a tree, then a corporation should cut about one-sixtieth of its trees each year. The problem in Montana also hinged on what is known as the "checkerboard," the apt description of an ownership pattern that gives private timber companies alternate square-mile sections of land.

BECAUSE THE NATIVE PEOPLE, PARTICULARLY THE LOCAL SALISH, could not conceive of owning land, there was no checkerboard in September of 1805 when Meriwether Lewis and William Clark made their way up the twenty-five-mile-long trough that holds Lolo Creek. In his journals, Clark only reports encountering hot springs toward the head of the drainage and his vexation with some devilishly steep land that left "Party and horses much fatigued."

"One Deer & Some Pheasants killed this morning, I shot 4 Pheasents of the Common Kind except the tale was

black. The road over the last mountain was thick Steep &
Stoney as usual."

Stoney, steep and thick with trees, the three causes of Lolo
Creek's subsequent undoing. In the late nineteenth century
the federal government began doling out huge hunks of the
wooded West, first as railroad grants to the Northern Pacific
Railroad, then 160 acres at a time to homesteaders who
claimed land under the Timber and Stone Act. Most of the
"homesteaders" actually were dummy entrymen hired by
timber companies. The lands coagulated as huge corporate
holdings and so the checkerboard was born. Linear thought
sent surveyors to dice the West into square-mile sections.
Since the government ceded alternate sections, color-coded
maps of western Montana look like a checkerboard. In the
Lolo Creek drainage, the squares are the green of the Forest
Service, orange for Champion, and lavender for the sections
held by Plum Creek. They are virtually the only players in the
fifteen-by-twenty-five-mile swath of land drained by the
creek, bounded on the west by the Idaho state line and on the
east by the town of Lolo, which is just about ten miles south
of Missoula. The Forest Service owns about half of that land,
while Champion holds forty percent and Plum Creek the
other ten. It is a drainage typical of the hundreds that finger
out from western Montana's major river valleys.

Champion and its predecessor, the Anaconda Copper
Mining Company, had controlled those lands more or less
for a century but had never logged them hard. Then in the
late 1970s, change came. Champion, a timber giant with
land and mills spread throughout the world, was pressed for
capital to build mills in Michigan and Texas. Lumber prices
were depressed. Buying logs from federal lands to supple-
ment the supply from Champion's own lands, a long-stan-
dard practice, would eat into the profit margin. So would
the use of environmentally sound but costlier logging methods. All this pushed toward a decision in Stamford, Con-

necticut, Champion's corporate headquarters, a decision that began to show on Montana's land in the late 1970s. Quietly the company decided it would no longer log its own lands only as fast as the lands could grow new trees. Instead, the company decided to log all of its lands as rapidly and cheaply as possible. My source said Plum Creek, at about the same time, decided to do about the same thing, meaning both would soon run out of trees. Because the cut from corporate lands was supplying about half the logs in Montana, the situation raised the specter of timber shortages. Further, the rules of the checkerboard compounded the effects of those seemingly isolated corporate decisions. Those rules gave the two timber corporations a substantial lever to use first against smaller competitors and ultimately to undermine national environmental laws designed to protect the forests and the creatures that depend on them. Lolo Creek was the first place those rules came into play, a situation that in miniature both explains and presages a plague that spread throughout the Northwest.

THE U.S. FOREST SERVICE IS NOTHING IF NOT AN AGENCY hell-bent on the cutting of trees. The agency is the prime habitat of the timber beast, and yet in 1987, Orville Daniels, supervisor of the Lolo National Forest, placed a moratorium on logging of all federal land in the upper reaches of Lolo Creek. His action was a frank admission that serious damage had been done. No longer thick with trees, those steep, stoney slopes that fatigued Lewis' and Clark's horses were beginning to slip away. The stone is granite, which makes a solid rock but a notoriously ephemeral soil, friable, grainy, and mobile before spring's rush of melt and rain. Logging on these rocky soils yields erosion. Studies showed that sediment from logging roads and skid trails was significantly harming trout populations in Lolo Creek. Further, Champion and Plum Creek had cut so many trees that the drainage's elk herd had no place to hide from hunters and from the even more threatening attentions of a mountain winter's winds.

Daniels's decision to in effect lock up federal lands for a decade rested on a provision of the formal plan governing the forest. The plan obligates the federal government to compensate for environmental damage on adjacent private land by not logging federal lands. It is that plan that is the corporations' lever, the jaws of the vise. While the two companies merrily buzz through the trees on their own lands, the feds have no choice but to respond as Daniels did. That dries up the supply of logs available to competitors. Once the supply of corporate logs is exhausted, however, Champion and Plum Creek will need federal logs to run their mills. These companies, unlike the smaller mills, have political bargaining chips to obtain them. More than eight thousand people work in the Montana timber industry, a healthy chunk of employment in a state with a population of about eight-hundred thousand. Historically, when the corporations have argued that they must have logs or close mills, they have gotten the logs. Further, Daniels set a ten-year moratorium, which means that once the corporations begin to need the federal logs, the competition for them will have been beaten. The independent mills will be out of business, and the moratorium will have expired.

Daniels and others in the Forest Service are blunt about this whole business. Lolo Creek's moratorium was only the beginning. When Daniels imposed it, Forest Service officials quietly predicted that the moratorium would spread, mostly because virtually identical circumstances exist throughout the western end of the state. (By 1991, that prediction appeared correct as the Forest Service began laying the legal groundwork for a much larger web of moratoria.)

All of this was only hypothesis in the spring of 1988. As far as the general public knew, both Champion and Plum Creek were still practicing sustained-yield forestry. The mills were churning out lumber at record rates and a steady supply of logs would roll on forever, we were told. In its inner circles, however, Champion acknowledged it had abandoned sustained yield in favor of what came to be known as the "accelerated harvest." But even on the inside, company officials said it would take at least thirty years to exhaust the supply of corporate logs. Because the program didn't begin until the late seventies, the shortages shouldn't emerge until a decade into the next century. In reality, the situation was far tighter, and Champion knew it. That was the fact I needed to prove and print, the heart of the story.

Armed with the broad outline of the details as supplied by my source, I began working the story. My first task was to locate harvest statistics independent of the two corporations. I had covered county courthouses long enough to understand that companies share a great deal of information with tax assessors, information that is public record. I asked some questions. As it turned out, a state tax official named Randy Piearson was building a study of taxation of forest lands and in the process had gathered data on Champion's inventory for every county in the state. (Piearson eventually learned that the average timber company pays total taxes of about fifty-six cents an acre on its lands. Champion pays less on a couple of square miles of trees than one of its workers pays on his three-bedroom home.) Because corporate property taxes in Montana are based on inventory, it is in the company's best interest to report the cutting of trees. Once land has been logged, taxes drop.

Piearson's data was still raw, but he agreed to assemble some of it for me on his own time. I met him on a Saturday afternoon at his house and took the numbers, went back to my own place and began punching them into my computer. It took most of the afternoon to enter the county-by-county data, but only a second or so for my Macintosh to spit forth a shocking figure. Champion's property-tax records showed it had already logged all but one percent of its merchantable trees. The numbers offered only two possibilities: either the company was fudging its inventory to cheat on its taxes, or it had cut in less than a decade the lands it said would take thirty years to harvest. When I laid those alternatives out for Champion officials later, they did not equivocate. The inventory figures were correct. Their lands had been cut.

The figures I had gathered through the tax records, however, only applied to Champion and so only accounted for about half of the corporate lands in the state. Curiously, for reasons never revealed but presumably because taxes were low to begin with, Plum Creek was not nearly so aggressive in seeking tax reductions. The company had not reported its cut for years and so left no independent paper trail. So far, I had only my source's word that Plum Creek was cutting its trees faster than they would grow back. That wasn't good enough to print.

"BILL PARSON IS A LYING SON OF A BITCH." THAT QUOTE WAS delivered on the record by Tom France, staff attorney for the National Wildlife Federation and Montana's most active environmental lawyer. As such, he has no interest in gratuitously inflaming the other side. Bill Parson, Plum Creek's chief executive for the Northern Rockies, is nothing if not the other side. His manifestos defending his business are Reaganesque and blunt, battering rams with the bark still on them. In the environmental community the majority holds with France that at least some of Parson's homilies do not cleave wholly to the truth.

One incident is legend. On a tour of Plum Creek land, Parson once stood before a group of reporters and environmentalists in a clearcut to deny that the company ever logged the banks of streams. He was standing in front of a logged stream. When one hopping-mad environmentalist pointed this out, Parson continued to deny the practice by saying the stream wasn't technically and in the strictest sense a stream. My head, however, stores another image of Parson.

A few months after the paper had printed my timber stories, the state's Environmental Quality Council, an arm of the Legislature, called a meeting on the issue. I was dispatched to the state capitol to cover it. Parson, who was there to testify, spotted me in the hall. He stomped straight toward me in high dudgeon provoked by what I had written. "No, more interviews, that's it." Then as he was skulking off, he turned to add, "Nice pictures though."

Missoulian photographer Michael Gallacher, my partner on this assignment, had shot the photos in Parson's office. It has a large picture window that overlooks Plum Creek's main mill yard. Behind Parson, the window frames the effluvia of industrial logging, steam venting from stacks, city-block sized piles of logs, pipes, mud, and sawdust. In the foreground, an open-collared checked shirt sets off Parson's sure blue eyes and military-cropped gray hair. Looking in charge and direct, he seems wholly carved from his background, and so he is. A man who is attentive to images above all else, he is comfortable when inset in a raw shot of industry.

I met him for the first time the day the photos were snapped. Gallacher and I had driven to Columbia Falls, a logging town on the west edge of Glacier National Park. We were early and tired from the three-hour drive through drizzle from Missoula. We sat in a coffee shop, watched log trucks roll by, and then talked with a state senator I knew who was running for governor. He had stopped in the cafe more for coffee than campaigning. In Montana, politicians still do that.

As we drove the service road to Parson's office, I was nervous. I needed something quite simple from Parson, an admission that his company had stepped up its cut beyond sustained yield. I had no way of confirming it otherwise. I needed numbers from a fellow who had always been far more interested in images, in reciting the gospel according to timber.

What I got from Parson that day was several variations of the toilet-paper gambit. It is one of the favorites of the timber industry, the contention that their sole aim is to serve obediently the public good by manufacturing products vital to American life, liberty, etc. Lately, industry spokesmen have been overfond of raising a haunting specter: should the environmentalists prevail, the fatherland faces an imperiled future without toilet paper. The argument has become the timber industry's corollary to wrapping one's self in the flag.

"Should the supply of forest products start to dry up or should the price go above what people are accustomed to paying, then . . . the public is going to start to put pressure on the public land managers to make that timber resource available so that they can continue to enjoy those nice forest products. Toilet paper is a classic example," Parson told me.

Plum Creek, however, does not cut its trees to sell toilet paper. Plum Creek's trees make chip board, studs, and planks or, increasingly, the company's trees make nothing at all, at least not in this country. This issue of log exports, Parson did not wish to discuss. When not sacrificing its trees in service of our backsides, the industry claims it acts only to preserve the jobs of millworkers. Never mind that automation has trimmed more than twenty percent of the timber jobs in Montana in the past decade. The contention of benevolence toward millworkers, however, flies flat against the reality that record numbers of unmilled logs are leaving the region for the Orient. Oregon, Washington, Idaho, and Montana now send each year close to four billion board feet of logs with the bark still on them to points east, mostly Japan and South Korea. And China, which was deforested a millennium ago. Four billion board feet is more than triple the total number of logs cut annually in Montana on all lands, state, federal, private, and industrial. One in five of all logs cut in the four states of the Northwest is exported raw.

With only minor exceptions, export of logs from federal and state lands is against the law. That means all of this trade comes from private lands. Plum Creek and Weyerhaeuser, the two largest private timber holders in the Northwest, also lead in exports. Plum Creek's response to this is to label it an issue not worth mentioning in Montana, because all of its export logs come from company land in Oregon and Washington. Montana logs are milled in Columbia Falls and other regional facilities. However, in a region with interwoven networks of trade, where log-starved mills from Oregon and Washington come to Montana seeking federal timber, state boundaries are meaningless abstractions.

What Parson deemed worthy of mention in the interview was a manifest failing of his firm that he wished to confess for the record. The problem, he allowed, was one of public relations: "We haven't done a very good job as an industry . . . in explaining the fact, in getting people to buy on with the fact that if they want these forest products at reasonable prices they are going to have to kill some trees."

Later, Parson told John Mitchell, a writer for *Audubon* magazine, that the sole problem was: "Our industry has not done a good job of telling our story to the public."

Despite Parson's fondness for this notion, it likely is not original. His boss, Dave Leland, Plum Creek's president and chief executive officer, later told the *Wall Street Journal*: "There's no question. We can't deny the fact that we have a bad image. . . . Have *had* a bad image." Leland said the solution to this lies in "more professional publicity work," a better job of letting people know "what we're doing and how we are going to do it."

In the world of Plum-Creek-speak, problems are all in handling spin. Having warmed to this theme, Parson brought it to bear on the key question of the interview: whether his company, like Champion, had abandoned sustained yield in the early eighties in favor of the accelerated harvest. He said, no, it hadn't, and although numbers were not right at hand, the company had cut roughly equal amounts of its own timber in each year through the past decade and a half. I asked him whether numbers could be at hand. He promised to send them later and then allowed that while the problem may not be Plum Creek's, it well may be Champion's, and further was not a problem of the industry over-cutting its lands but rather a problem of the Forest Service undercutting public lands.

"In the future, maybe twenty years from now, the public timber managers . . . are going to have a choice. They're either going to have to start harvesting their timber and carrying their proportionate share of the load, or they are going to have to to convince two-hundred-twenty million Americans that they can't have the forest products they have traditionally enjoyed at the prices they have traditionally enjoyed."

Throughout the latter half of the eighties, about half of the timber cut in Montana was coming from Plum Creek and Champion lands, half the harvest from about eleven percent of the timbered lands. Industry does not derive that much timber from such a small area unless it squeezes hard. Parson argued that the real problem is that the public's land is not hit hard enough.

I asked him about allegations that his company had used shoddy and damaging practices.

"I don't think we've done anything wrong in that area. The prescriptions we have applied on the ground have been sound silvicultural prescriptions," he said. "We intend to do the right thing. That's not a Bill Parson goal. That's a company goal."

End of interview, but Parson did not want me to leave. He wanted to make sure I saw the company nursery in a neighboring town where its next generation of seedlings was being incubated. And he wanted me to see the company's PR videotape showing happy workers, wildlife, and vigorous saplings arching to the sun. Gallacher and I sat through the tape in his conference room. Parson watched too, then left me with a curious statement.

"It may not be the way we are," he said of the tape, "but it's the way we would like to be."

He expected that I would not write of things as they stood, but as Bill Parson would like them to be. He wanted me to accept the same set of images that kept him coming to work each day. Images, however, would not support the story I was writing. It needed confirmation that Plum Creek had accelerated its harvest, but leaving the interview, I had only Parson's vague denial that the acceleration existed. The story still had a hole, but a few weeks later I received a letter signed by Parson himself. It contained the numbers I had asked for, that he had promised but that nothing other than his word compelled him to give me. Between 1981 and 1982, Plum Creek had doubled its harvest. By 1987, it had tripled the harvest rate of the seventies, a level it continues. Those numbers flatly contradicted what Parson had said in the interview. My story was back in business. For at least six years and to the present, Plum Creek had foresaken sustained yield to cut the hell out of its trees. I have no idea why Parson decided to confirm this.

FOR MORE THAN A CENTURY, 1.4 MILLION ACRES OF NORTHwest forest basked relatively unmolested, the beneficiary of Plum Creek's obscurity within the network of businesses commanded by the Burlington Northern Railroad. Those lands were the dowry awarded the railroads in the marriage of Midwest to Northwest, a carrot extended by President Lincoln in 1864. In turn for accepting the risks of building railroads, the government ceded the railroads vast tracts of federal lands. Corporate predecessors and then BN itself used the land for a few railroad ties, some corporate retreats here and there, lodges to entice passengers, but the business at hand was railroading. Ignored, the trees grew, became timber, and had swelled to a looming reservoir even BN could not ignore in those heady days of the eighties, when Reaganomics began to bloom full force and America was in business again. It was the day of the MBAs, economic Darwinists, nut-cutters. This mindset fostered, from the forests' point of view, a fatal abstraction: trees became inventory, worse, slow-growing inventory susceptible to fire and bugs.

Said Parson of the stretch of the natural world in his charge: "If you had that kind of investment in your portfolio, you'd get rid of it and get something that was growing."

More particularly, though, Plum Creek officials now acknowledge that a form of predation in the corporate habitat was largely responsible for accelerated harvest. The eighties was an era of takeovers. Firms with undervalued assets like timber were particularly vulnerable prey. The solution was to cut and sell the timber as rapidly as possible, converting the asset to cash, reinvest it, and hold the asset strippers at bay. Plum Creek took that solution.

In the early eighties, Dave Leland took over as Plum Creek's president and chief executive officer. A sleepy subsidiary of BN woke up. Leland straightway streamlined the firm by "downsizing" its staff, particularly foresters whose job it is to ensure that logging operations are silviculturally and environmentally sound. At the same time it doubled its cut. Parson acknowledges that thirty-five foresters in his employ got the ax.

One of my sources, a Plum Creek forester in the early eighties, said the company's attitude shifted dramatically. Foresters who protested the heavy-handed practices were given a clear choice: "They would tell you either to do the job or someone else will," this source said. "The joke was you would come in from the field and your desk would be gone, but it wasn't a joke, it happened . . . . I think the way the company looked at it was they saw all these trees, all this standing old-growth timber, as money in the bank."

In 1989, BN pulled off a couple of neat corporate splits that eventually landed Plum Creek clear of its parent company as a limited partnership, culminating a series of moves that company officials acknowledge left shareholders "several billion dollars" richer. By 1990, Plum Creek was openly referred to as "the Darth Vader of the industry."

In 1990, two years after Bill Parson ducked my questions about sustained yield, the company admitted cutting its trees twice as fast as they would grow back. Parson told a Montana state legislator, "We have never said we were on a sustained-yield program, and we have never been on a sustained-yield program. Let's get to the heart of it. Sure it [corporate land] is extensively logged, but what is wrong with that?"

WHEN GALLACHER AND I LEFT PARSON THAT DAY, HE TOLD US we really ought to visit the company's nursery in a small town about halfway between Missoula and Columbia Falls. There the company incubates the seedlings, rows of them, straight, uniform, and with genetics certified to warm an accountant's heart. This nursery, the company says, is an emblem of its good intentions. Gallacher and I had seen nurseries before. The story would not dispute the fact that both Champion and Plum Creek do plant trees. It's in their interest. There is a federal tax break for reforestation. The real issue is whether those trees will grow, or more complicated still, whether one can replace an entire natural community we call a forest by planting only trees. A simple tour of a nursery could reveal nothing about that issue. We declined Parson's invitation. Instead, we chose to visit another piece of Plum Creek property, hard-cut timberlands in a place called Jim Lakes.

The lakes lie in alpine country, high country that rolls against the Mission Mountain Wilderness Area. The east slope of the Missions is fingered with a series of canyons that slide creeks down to either the Swan or Clearwater rivers below. At the head of each drainage runs the wilderness boundary, federally protected wilderness lands where machines, including simple ones like bicycles, are banned. Against these boundaries, however, lie sections of Plum Creek land, and so one can walk this strangely symbolic and still starkly real line where wilderness and civilization collide. Jim Lakes—there are two of them pooled at the center of a bowl of land ceded to Plum Creek—lie within sight of this line, or at least in sight now that the trees are gone. But it's best to visit Jim Lakes by first going to Cold Lake, a mile-long bowl at the head of a parallel drainage, parallel in elevation, slope, climate, size, and we can now only imagine, once parallel in vegetation. The difference is, Cold Lake is inside the wilderness boundary. You walk there up a middling steep swath of forest trail canopied by bearded old spruce, lodgepole, and subalpine fir.

At the lake there is a campsite, unused. It has been circled with a ring of twine and a small notice from the Forest Service asking hikers not to camp here for a year or so. The soils and plants of this alpine terrain are notoriously fragile. They will withstand not even an occasional crushing by a nylon tent, not even the thump of a hiking boot's sole. This kitchen-sized chunk of tender land needs a rest.

Just a mile or so away at Jim Lakes there is no rest. When I was there the clearcuts were but a couple of years old, a rocky, mountain amphitheater stripped of all trees. These alpine lakes and streams are able to hold life only to the extent that a ring of trees can shield them from winter winds and intense mountain sun. Because of this, forestry standards that Plum Creek itself has endorsed and claims it enforces forbid logging the edges of lakes and streams. The banks of Jim Lakes and the creek that veins them are bare. Near the outlet of one lake, we saw gaping bulldozer bites out of each side of the creek's bank, cleat marks where the cat clomped straight across the creek. This is a violation of forestry standards. This cat cut stands just a few feet downstream from a bridge. Presumably, the cat's operator believed the wooden bridge would not suffer the weight of his machine, so the creek did.

Across fresh-built roads, fans of eroding sand wash down slope, ultimately toward Jim Creek, a creek that used to raise young trout until the sand filled the space between the rocks where fish usually lay eggs.

None of this damage derives necessarily from the simple act of cutting trees. These scars on the land are rather the result of how one cuts trees, the issue of forest practices. There is a band of practices, ranging on its gentlest hand to the snatching of an occasional tree from a dense stand and snaking it out with a horse all the way to the evidence lying on the land around Jim Lakes. This is forestry at its harshest, not really forestry at all, but more a form of strip mining. At its most severe, this sort of cutting proceeds across very large clearcuts, hundreds of acres at a time stripped not only of mature, usable trees, but of all trees, all vegetation. The sawable is sawed, the marginal is burned, the limbs and brush are burned, the land is burned and then a few years later crews plug in genetically and economically acceptable saplings. A forest is reduced to the mountain's equivalent of a Midwestern cornfield in a massive ecological and genetic experiment. A couple of hundred years from now, we will know if it worked. If it doesn't we will know the corporations got the last of the wood out fast and cheap.

Arnold Bolle has a name for this. Bolle is a softspoken old man, dean emeritus of the University of Montana's School of Forestry. Likely he had a hand in the training of more foresters than any person in the West.

"It's Nazi forestry. You clear off all that old junk and put in a good tree of good genetic quality in orderly rows as if that's the whole reason God created trees, just for our benefit," he once told me. "It's a very comfortable thing to think that man is in total control and everything is obeying us."

Yet in a legal sense, asserting that "man" is in control of Plum Creek's land overstates the case. In Montana, Plum Creek is in control, largely because Montana, unlike all other timber producing states in the Northwest, has no law governing forestry practices on private lands. Despite the fact that these practices affect wildlife, water quality, and irrigators well beyond the corporate boundaries, Plum Creek may do largely as it will on the land the federal government ceded to it. Not that people haven't tried to gain some control. Repeatedly, Ben Cohen, a state representative who once worked for the Forest Service, has introduced bills to create a forest practices act, but it is a measure of the political clout of the industry that those attempts generally die in committee. Part of the credit for this goes to the state's top forester, who publicly counsels that the state doesn't need a forest practices act. Privately, he has offered a different opinion.

Technically, Gary Brown's job is to manage all state-owned timber lands, a job that gives him the title of state forester. When the state, however, began toying with the notion of a forest practices act, Brown became the official point of concentration for legislative investigations of activities on all forested lands. His advice to the legislature was that it should not pass a law mandating sound forest practices. Brown's preferred method was to jawbone the corporations into voluntary compliance with an industry-drawn list of prescriptions called "best management practices."

Not long after I spoke with Parson and saw his company's handiwork on the land, I interviewed Brown in his office to see what he thought ought to be done about shoddy forestry in the state. That, he told me, was "old news." I was only focusing on the negative, out to give the craft of forestry a bad name at the same time that the industry had cleaned up its act. He became visibly angry at one point.

"I guess I don't like the way this conversation is going," he said.

For the record, Gary Brown felt all was well on the land and that the state shouldn't gum up the works with a new set of regulations. For the record.

In my boxes of notes, I have transcripts of interviews completed by another reporter. In these interviews sources inside the state forestry office described practices similar to those I saw on Plum Creek lands. Those sources spoke of irreparable damage by bulldozers on steep slopes, of erosion and the subsequent loss of topsoil that makes the regrowth of the next generation of trees difficult if not impossible. They even told a story I heard subsequently from other sources about a committee the state set up to head off some of the problems in checkerboarded drainages. State, federal, and industrial managers would cooperatively schedule cuts in a given drainage to spread the shock, limit damages to water quality, and rotate the harvest fairly. Industry, however, used the information to anticipate where the public entities were headed next and beat the state and federal loggers to the punch.

One source said:

> I heard the other day . . . we provide our six-year cutting information to all the members of the co-op and some of the guys up in Kalispell feel that all we're doing is advertising where we're going to be moving into in the future, and the industry is moving in ahead of us, because they've got the information and they know where we are going to go.

Under the rules of the checkerboard, the feds must mitigate damage caused by cutting on private land by choosing not to cut neighboring public land, and so the players stand at checkmate.

The source in the state forester's office commented on the hottest issue, the need for a law setting forest practices.

> We are the focal point for forestry in this state . . . Therefore, I hope in this process, and this is off the record, . . . I hope we end up with a forest practices

> act, something that is realistic so that we can carry
> it out, because we don't have a mandate right now.

Partly because of Gary Brown's subsequent public pronouncements, the state forester does not yet have such a man-
date. Plum Creek and Champion still may behave as they
wish on timbered lands in Montana.

BECAUSE I TOOK NO INFORMATION FROM THE MAN I AM ABOUT
to describe, he was really more a guide than a source. He
knew the logging roads and clearcuts of the Seeley-Swan
Valley. There, people work cutting trees, as any view from
atop the ranges that rim the valley will announce. From elevation's vantage, one sees little but a web of clearcuts tiling
the valley floor below.

These patches, some a mile square, are mostly the work of
Plum Creek, which owns hundreds of sections of land in the
twenty-mile wide swath between the Bob Marshall and Mission Mountain Wilderness areas. Logging carves a hard face
on the land, on the people, and the towns. Shops and cafes
wear day-glow green signs announcing their sympathies for
the timber industry. Environmentalists are called "tree-hug-
gers" and "flower-sniffers." The local bank sends out pro-
logging tracts with its monthly statements. Environmentalists
and industry tussle over the right to screen their respective
video-taped propaganda to elementary school classes. Four-
wheel-drive pickup trucks, hauling rows of chainsaws
stabbed in racks like oily Excaliburs, nose into a string of tav-
erns at quitting time.

Inside, men uniform in wide, red suspenders pursue their
leisure with the earned abandon common to any working-
man's bar, but the mood can blanche quickly. Eyes and voices
harden when people speak of timber, which is why my guide
wished to remain anonymous. He lived in this valley and
drank in these bars, so I will call him only "the Savage." He

was much like the cedar savages I knew as a kid.

He was an environmentalist, I guess, but an odd variety. In his thirties, wiry and full of a cackling chatter, he made his living catching what fell through the cracks, scavenging a bit of firewood here, some cedar shakes there, probably killing an occasional out-of-season deer. He knew in great detail how to shoot the locks off Forest Service gates that closed old logging roads. He was full of racist theories as to why corporations were clearcutting his valley. For this I thought him an idiot. I thought further, though, when I learned that he once pushed a paraplegic friend across about forty miles worth of the Bob Marshall Wilderness. The pair—the crippled man in a handcart—covered about two miles a day of mountainous foot-trails. This friend was ill, and the Savage believed one should not die without seeing the Bob.

But on the day I am remembering, the Savage was guiding Gallacher and me to some of the more egregious logging sites of the Seeley-Swan. It was the day we found Jim Lakes. Beyond the lakes, though, we saw more clearcuts fingered with erosion, stream beds stripped, lakeshores trimmed bare of trees, and mile on mile of permanent roads bleeding their sediment into nearby trout streams. These are the marks of logging. We were driving those roads in my Jeep, but sometimes we'd walk, for a feel of the place. It was a hot day, and yet in some uncut places we walked beneath middle-aged spruce, cedar, and fir, stepping on the cool, sponge floor of a live forest. Then the forest broke to a clearcut. The temperature jumped twenty degrees. The air lost its load of chill vapor and the ground rattled hard and dry beneath our boots.

We walked in one spot where a logger worked. We talked to him after he used a crane-like log loader and an aggressive dog to block the road that was our exit. He shouted at us over the pocka-pocka of his idling diesel. We told him what we were doing; he did not seem to appreciate our work. We read the bumper sticker on his truck announcing that his property was "insured by Smith and Wesson." He glared from the loader's seat and explained how environmentalists stood in the way of the huge machine's huge payments. Were we environmentalists? We guessed we weren't. He moved his crane and dog to let us go.

A few hundred yards down the road, the Savage made us stop the truck so he could rail a bit. Plum Creek only wanted a few of the species that were growing in that clearcut. The remaining trees, at least a quarter of them, the loggers had bulldozed into piles the size of houses. These logs they wouldn't saw, but burn. The company, however, believed firewood scavengers might be injured and file lawsuits, so forbade the collecting of firewood. This angered the Savage as he leaped around a pile of logs large enough to heat several houses for several winters in a valley where poor and rich alike heat with wood. He showed us dead buried trees a couple of feet in diameter and said they would have made decent cedar shakes.

Plum Creek was not interested in harvesting cedar. In that valley, it was not considered a "commercial" species. Plum Creek was more interested in stripping a forest of trees it considered undesirable so that it might plant species more usable in its mills. This is how Plum Creek farms trees. Throughout the industry, such a practice is considered sound silviculture. These are the rules of the land.

We left that cut, wound down some more logging roads, then up a long grade toward the Alp-like Mission mountains. We drove onto one clearcut of about forty acres, spooked a small grizzly bear that had been foraging at the edge of a seep. The bear—it looked to be a yearling—lit out for cover waddling like a fat kid in a fur coat. Gallacher swore, leaped for his camera gear, and missed the shot. The Savage lit a joint and allowed as to how the experience was "far out." Word choice notwithstanding, I agreed with the Savage. You can live in these mountains a lifetime and never see a grizzly, although this is their last remaining habitat in the lower forty-eight states. You can't live here long, though, and not want to see one. They are the essence of the wild. Some of us who love these Northern Rockies hold grizzlies to be harbingers of great mystery.

CHAPTER THREE

IN A MONTANA mountain valley called Gold Creek
there is a grave, a life laid down in the soil. Likely as not a pioneer woman named Mahala Jane Primm is buried here,
alone in the middle of the woods, because in her day that's
the way it was done, you died and whoever was left buried
you on the home place. Likely as not, but there is evidence to
suggest otherwise, evidence to suggest she chose this spot because this woman cared for trees. Looming evidence. Behemoths of ponderosa pine, hundreds of them, five feet across
at the base, more than one hundred feet tall, more than five
hundred years old, trees whose roots clutch the same soil that
once held the roots of this woman's life. In this spot is peace
guarded by the ancient wisdom of the forest.

This grave anchors but two hundred acres of sanctuary in
a county-sized valley where the main business is cutting trees.
Around the grave and beneath the ancient trees the soil that
holds this woman's death is quiet, but breathing, balanced
and alive, humming with the rhythms that drove her life and
have driven countless lives in cycles since the glaciers let go of
this land ten thousand years ago. A few hundred yards away
begins the cutting of trees, hundreds of square miles where

not just the trees, but the soil that raised them, is beaten and banished, a casualty of a fierce battle. This woman spent forty of her seventy-nine years here, then died on her land. These trees have lasted five centuries and will spend another two hundred years in their falling and rotting. Yet this soil that made woman and trees was at the very least seventy centuries in its own making. The story of people and trees begins in telling about dirt.

The forests of Montana cling to mountain slopes only because of monumental caprice and catastrophe. That trees stand here at all is an accident, one that is easily reversed. The odds are against trees growing here, because the land is dry and hard. Tectonic plates warped and folded the Rocky Mountains really not so long ago, and so the emphasis still is on "rocky." The glaciers are barely gone in most places, linger still in some. Ice-ground gravel lies beneath but a skiff of dirt on the moraines that rim our valleys. In most drainages, very little rain falls. When it does, it bolts down slope to mountain streams that spirit it away, or slips through gravel subsoil to aquifers well beyond the grasp of tree roots.

These circumstances do not welcome trees, but a geologic accident counterbalanced the harshness. About seven thousand years ago Mount Mazama erupted, tipping the scale in favor of trees throughout the Northwest and especially here in Montana. When Mount St. Helens in Washington erupted in 1980, western Montana got but a dusting of ash. When Mount Mazama exploded, it turned a ninety-nine-hundred-foot mountain into the six-mile-wide hole that is Crater Lake in western Oregon. The ancient eruption was farther away from Montana than Mount St. Helens was, yet ash piled up here by the foot.

To western Montana, this cataclysm was creation. That layer of volcanic ash was rich in nutrients, and it could hold water. It could grow trees. Wispy as it was, it still stuck long enough to give pine, fir, and larch a toehold. The trees in turn sheltered that ash layer with the results of their own lives, a blanket of dead and detritus, and so the life of the soil began with the laying of lives in the soil. In this decaying organic stuff that still shelters the ash that remains, bacteria and fungi prospered, completing the circle that makes soil live. Microorganisms break down minerals from ash and rock to feed trees that die to feed microorganisms.

The cycle of soil glues our rocky mountains to life as these mountains know it. As vital as that cycle is, you would think we would understand more about it. We don't. This piece of the story is about the collision between the delicacy of soil and the cleats and blades of bulldozers, the same collision that sums the advance of our people west across the soil of the Great Plains and into the timber and mines of these mountains. That sweep of history, however, played and plays with particular severity here in western Montana. Volcanoes shaped this land, but so did people, and now, so do their bulldozers.

WHILE VOLCANOES, ROCKS, AND ICE BROUGHT THE SOIL, PEOple have been here even longer than the dirt. Most of these people, however, managed without bulldozers. Evidence suggests these mountains knew human tracks even before the glaciers were fully gone, for at least fifteen thousand years, but for one hundred forty-nine of those one hundred fifty centuries, not much changed in the life of these trees. That's not to say people didn't use them. Ancient mortars and pestles turn up from time to time among some forests near timberline, suggesting that the nuts of the whitebark pine there were an important source of food for prehistoric tribes, as they were and still are for grizzly bears. Bark strips made baskets. The reedy lodgepole pine made serviceable tipi poles, and so a few of those were felled to frame the lodges of the Blackfeet, Salish, Kootenai, and Nez Perce in knowable history and of the unknowable tribes that came before that.

There is evidence that it was the Indians who first began meddling on a grander scale. About the time that descendants of Spanish ponies began filtering in from the Southwest, the frequency of forest fires doubled. Apparently the tribes were burning underbrush to stimulate fresh, new grass for horses. Still, this was but a quickening of the pulse of a normal event. The Rockies are naturally tested by fire, mostly set by lightning. Fire trees grow here, especially ponderosa pine and Western larch. Both have thick, fissured, corklike bark that insulates their vital parts from flames. Fire thins the trees' competition, leaving evenly spread spars of ancient pine and larch over park-like grass savannahs and shrubs. Left to its own devices, a ponderosa can live six hundred years, a larch as long as nine hundred. Both will grow taller than one hundred fifty feet. The hills of western Montana still hold a few ancient trees. These trees have escaped the saws and bulldozers, but they have not escaped fire. Their trunks are scorched and scarred by fire every decade or so, as regular in the trees' sense of time as winter is in ours, and still these trees tower and live, as much a part of their place as the fire that tests them. If they are left to their own devices, they live on as they have since the Ice Age, at least until now.

The culture that colonized this place late in the nineteenth century had a notable eccentricity: a peculiar urge to own things, even when "things" included six-hundred-year-old trees, seven-thousand-year-old soil, and everything created by their conjunction. Over the years, about ten million acres of trees in Montana remained in the public domain, and that is hard enough on the trees. Hard enough, because the U.S. Forest Service, charged with looking after them, really exists to cut trees. It is run by the timber beasts. In recent years, the Forest Service has made noises about reform, has been forced to reform by a string of national laws inspired by the environmental movement and by the demand for more recreational use of the forests. To a degree, this has fostered some protection for that part of creation that does not fit on an accountant's ledger.

Still, the Forest Service keeps a ledger, thereon allocating by far the biggest hunk of its budget to the cutting of trees. This fact is kept in computers, but mostly it is written on the hills. Still, there are at least rules on Forest Service land. The Forest Service will now tolerate no clearcut larger than forty acres. I have seen clearcuts—the practice of stripping an area of every single tree—larger than a square mile on corporate land. The Forest Service has rules about where the cats can go. On private lands, cats roam anywhere there is money to be made. The private land is used hard. In 1986 in Montana, about half of all logs cut came from eleven percent of timbered lands, the property of Plum Creek Timber Company and Champion International. The two corporations own land that used to be public, but fell to private hands largely in two ways.

The first was governmental largesse to the railroads, which provided Plum Creek's acreage in Montana. Champion International owns about an equal number of the checkerboard's squares, but these lands have an even more checkered past that is rooted in theft. Actually, it would not be too difficult to defend this larceny. The only way in the middle part of the last century to obtain government timber (which was to say all timber in Montana) was to steal it. The nation's forest reserves were still ruled by an 1831 law that considered the government's trees a naval reserve and as such not cuttable or saleable in any way.

The law enraged westerners who viewed it as a cramped Eastern mindset, not at all in touch with the needs of a Western economy that was building fast from a seemingly endless stock of trees. In Montana, this interregional war was fueled not by the timber industry but by the copper mines at Butte. The mines needed timber to shore up their adits, fuel wood for their smelters, and lumber to build whole cities. By 1884, the Anaconda Copper Company at Butte required three-hundred thousand cords of firewood a year. Aside from firewood, it was using about forty thousand board feet of timber (enough to build four houses) each day. To meet this demand, Butte turned to Missoula, a hundred miles downstream on the Clark Fork River. Where the Big Blackfoot joins the Clark Fork, there arose a mill now owned by Champion, now one of the world's largest plywood mills. But in the days of the big trees, the product was not veneer, it was timber, and loggers ventured up the Big Blackfoot seeking the larch and the pine. They marched up the main river, then on up the tributaries like Gold Creek onto then-government land.

In that first skirmish of this war on trees, the sawyers and fellers were marshalled by a group of Missoula businessmen known as the Montana Improvement Company, from all reports a dedicated group of thieves. In 1885, a federal agent charged the new company with stealing forty-five million board feet of lumber and eighty-five thousand railroad ties from federal lands. Agent M. H. Haley labeled this theft "universal, flagrant, and limitless." He further alleged that the Montana Improvement Company was formed ". . . for the purpose of monopolizing timber traffic in Montana and Idaho, and under contract with the railroad co. running for 20 years, has exploited the timber from unsurveyed public lands for great distances along the line of said road, shipping the product of joint trespass and controlling the general market."

Haley's investigation triggered a high-stakes game of politics that echoes today. The businessmen said that if they couldn't steal federal timber, they would have to close the mills and throw ten thousand people out of work, an argument still made. T. F. Oakes, head of the Northern Pacific Railroad, which owned stock in the timber-stealing venture, wanted the unpleasant investigation to end. In a letter, Oakes leaned on Montana Gov. T. S. Hauser, who also happened to own the bank where Oakes's railroad kept most of its money. The letter said:

> Read this over carefully and let me know if you
> intend to take a position in reference to our timber
> interests. If we have no rights in this property you
> will respect, I shall at once draw our deposits from

your bank . . . and in every other respect make things so hot for you, you will think the devil is after you. The Northern Pacific Company has not spent $70,000,000 to be bulldozed by you or anybody else. Let me know what your position is. The Northern Pacific Company has the right to demand of you the fullest support in every reasonable effort to protect its interests. It has never asked anything of you thus far but has done a great deal for you and your interests thus far with very little return.

The blackmail paid off. The trial of the timber thieves stalled for twenty-six years, and then they were convicted on drastically reduced charges. Meanwhile, the thefts continued and the westerners scored a victory of sorts in the form of the Timber and Stone Act of 1892. That let the yeoman foresters of popular myth claim small parcels of forest land as their own, a sort of Homestead Act for the forest. It had, however, a loophole punched open by the raw-knuckled politics of the time. The timber barons used employees as front men, "dummy entry men," to assemble vast holdings of federal land at about $2.50 an acre. During this period, 663,552 acres of federal timber in Montana fell into private hands, which eventually came to be controlled by the Anaconda Copper Company, the hand that held everything in Montana through most of the last century and this one.

Homesteader Frank H. Parker, Mahala Primm's predecessor in Gold Creek, was one of the few who did not sell his land to "the company," as it was known. He held and worked his land, built cabins and barns, pastured cattle, and even cut a few trees. Parker was the exception to the rule throughout the Northwest. What began as a sort of Homestead Act left half of all the timber in the Northwest in the hands of thirty-seven holders. One quarter of this was held by three companies, the Southern Pacific Railroad, the

Northern Pacific Railroad and Weyerhaeuser, which still exists, and still ranks as the nation's leading exporter of unmilled logs to Japan.

In Montana, Anaconda Mining used its trees, built its mines and smelters and eventually built a network of political and economic control so tight it was known as the "copper collar." The stranglehold held fast until the late 1960s when the dynamics of the world copper market began to undermine Anaconda. In 1972, Anaconda's mine at Butte, one of the world's largest open-pit mines, closed. During the same year, Anaconda sold its timber lands to Champion.

Thirteen years earlier Anaconda sold its newspapers. It owned four, the dailies at Billings, Butte, Missoula, and Helena (the state's capital). The *Missoulian,* the paper I worked for, once wore the copper collar. In 1912, when the trials of the businessmen accused of timber theft were beginning in San Francisco, C. H. McLeod, one of the ringleaders of the theft ring and one of Missoula's most influential businessmen, contacted the publisher of the *Missoulian.* He asked the publisher to print reports of the trial favorable to the company. The *Missoulian* did.

Late in the twentieth century, I should have expected to trip across that entanglement of journalism and commerce once again, but at the time, I was not really interested in that land's history, human or otherwise. Webs of interdependence, webs such as politics, economics, and journalism, or even more organic webs such as trees and fungi and nutrient cycling, were not yet on my mind. I was more attuned to the sounds of bulldozers. By this point, I was a month or so buried in the story of Montana's logging industry, still working the story in a conventional manner. Because of this, I was more interested in—even obsessed with—the bulldozers. There were other parts of the story, but early on a soils expert had told me about the fundamental tale of the ties between trees and the soil. Other sources had told me the cats were doing great damage on the mountainsides. I wanted to find the cats and photograph them to document the damage first hand. One day, photographer Michael Gallacher and I did just that. This day produced a string of images that did more than anything else to shape the story. We came to refer to these photos, in the patois of post-Watergate journalism, as the "smoking gun."

WE WERE VISITING GOLD CREEK'S VALLEY, WHERE CHAMPION now owns and has mostly cut at least one-hundred square miles of land. There is very little checkerboarding in Gold Creek. Instead, it lies under a mostly contiguous blanket of Champion's ownership. It feeds the mill at Bonner. As we drove up that drainage, Gallacher and I became more and more excited, because it appeared we were going to find the photos that would document our story well. My Jeep had traveled a full fifteen miles up dusty logging roads, passing through devastated clearcuts the whole way. It was as if we had somehow left what we knew as Montana. The valley that holds Gold Creek is but ten air miles from Missoula, but it is strategically screened from the major highway nearby. There is no reason to drive into the drainage other than to cut trees. We and most Montanans never visit places like this, a sort of loggers' hell.

On previous trips, Gallacher and I had seen mile on mile of hard-cut land, yet nothing like this, a once-forested mountain valley worked as hard as a strip mine. Champion had "slicked off" (the local term for a clearcut, one the loggers themselves use) its land in other areas of the state, but most of its land elsewhere is interspersed with federal, state, and private holdings. In this drainage, the contiguous ownership makes the evidence of the corporation's work roll out to the horizon like a tidal wave of deforestation.

As we drove, the cuts became progressively fresher, meaning we were nearing active logging sites. We could have found logging faster by accepting company tours, but for the obvious reason, we went about this business the hard way. We did not want to see what the corporations wanted us to see, so we searched. That involved driving a few miles then stopping and listening for the whine of chainsaws on the spring air.

After a couple of hours, we heard a saw clatter from a distant ridge and headed toward it. A logger named Kevin Rausch was reworking a seed-tree cut, a common logging technique. On the first pass, loggers all but clearcut an area, leaving standing only a single mature Western larch on every half-acre or so. That produces a forest covered with trees about as thickly as goalposts cover a football field. After those parent trees have shed a few years' worth of cones to seed the next generation, loggers cut them down, which Rausch was doing. He worked on a ridgetop commanding a view that summed the recent history of Gold Creek. Stretching to the sky behind him were mile on mile of bald slopes webbed by steep gravel roads for logging trucks and fingered by the trails of the cats that had skidded the trees.

Before now, I had seen this spot only on computer. I had crunched through the state tax records which revealed that Champion had cut most of its trees. Yet I had no feel for that statistic until I saw it printed on the land by the mile. Officially, foresters from both Plum Creek and Champion had told me such scenes did not exist, that, yes, they were cutting, but they were using techniques to ensure regrowth of the forests and to prevent erosion. Yet erosion and tactics that cause erosion were everywhere. Topsoil was visibly in flight. True, the next generation of trees was being seeded, but how could it grow when the integrity of the soil was so undermined? This was not a question I had to form. This was a question as clearly visible on the land as the rutted skid trails that raised it.

When we spotted Rausch, Gallacher understood immediately that he was looking at an image that told the complete story. He bolted from the Jeep and began loping straight up the hill, hurdling stumps and brush as he went. Gallacher is in his late thirties and still in solid physical shape. He has to be to support his constant frenetic pace. He exists mostly as a ball of energy. Still, the scene before us made him a bit more frantic than usual as he vaulted up the hill like a large ape.

He got the photo he was after: a single frame shows a background of skid trails and clearcuts, with Rausch—suspenders, grimy pants, hard hat, logger boots, and Swedish chain saw—collected in the foreground of the 35-mm frame. A middle-aged larch cut clear through at the stump is falling, frozen by the shutter as it leans to the right and off the frame. The scene shows little but stumps made and stumps in the making, and both of us, on seeing it, felt that split of loyalties that eventually plagues most journalists. We had captured the image we knew was bound for Page One. As reporters, we had done our jobs, but as humans, we wished we had never seen that place.

This schism between work and place showed up repeatedly, not just in us, but in those we met on this story. Rausch shared it. In simpler times, he had learned a craft regarded as honorable. Now, though, he works in this new world of the endless clearcuts that make a sustainable partnership with the forest impossible. Rausch does not like his role. I am not guessing about this because I talked to him for a bit, the sort of contact that had intimidated me in the early days of the story. Gallacher and I were photographing the loggers and speaking to them amid scenes of such obvious environmental poundings that we at first figured the loggers would resent us. We were pointing a public and accusing finger at the way they made their living, and no one can like that.

Yet throughout the story, we found the loggers to be frank and approachable, willing to speak, but often resigned to the paradox of their lives. They did not like what they were doing, but it was the work they knew, and the work that at once bound them to and split them from these mountains. Rausch felt the bind as he answered my questions. He glanced grimly over the sweep of clearcuts behind him then said, "Maybe if we would do a better job, the environmentalists would get off our backs."

Rausch was telling us something important: this story was not so much about the cutting of trees as it was about the ways in which we cut trees. These days, our methods are all aimed at profit, or more precisely, profit in the next quarter as opposed to profit twenty years hence. In this taking of the forests for our own benefit, an act as old as humans, there are compromises to be made in the interest of ensuring a future. There are ways to cut that can ease the pain. In Montana, it was not news that industry was cutting trees, but what was news was that quietly, over nearly a decade and out of public view, the industry had responded to tough times by giving no quarter. The industry existed to get the wood out and get it whacked into studs, plywood, and cardboard boxes as fast and as cheaply as possible. That was news and a story the bulldozers showed.

WE FOUND THE DOZERS THAT DAY SIMPLY BY ASKING RAUSCH where they were. They wouldn't be skidding the trees he had cut for a few days, but Rausch said that over there, off toward the horizon, hidden in one of the draws that veined the drainage, we would find the cats. We thanked Rausch, climbed into the Jeep and bounced over more miles of logging roads, over ridges, around the switchbacks until we finally found the cats. They were exactly as my sources told me they would be, working slopes almost too steep to climb on foot. If that same ridge were on public land where rules govern forest practices, those cats would not have been there at all. Those slopes were far too steep for tractor skidding under federal guidelines. Instead, the Forest Service would have required a gentler aerial system of cables to snake logs upslope. Cats churn the topsoil and make it vulnerable to erosion, especially vulnerable here on slopes so precipitous. Yet cats are about half as expensive as any other method of skidding.

Gallacher and I left the Jeep and scrambled over stumps and brush to the top of a protected ridge were we could set up telephoto lenses. Gallacher's motor drives spun through roll after roll. I watched through binoculars. The cats worked. The cats started from a logging road at the toe of the slope and quartered their way toward the top of a clearcut so fresh that most of the shrubs and grasses of a sheltered forest floor were still standing and green. It was a bit like looking at the intact living room of a house after the roof and walls had been removed by a tornado.

At the top of the ridge, the cats pivoted and pointed their blades straight down slope along already-well-grooved trails. Immediately, they dropped their blades for a bite of soil, a method of slowing their descent. Brakes alone would not check the momentum of the behemoths on these steep slopes. As the cats neared piles of felled logs, the operators set the blades down hard to jam them to a full stop, scouring another bite of soil. The operator left the cat and pulled a cable called a choker from a series of such cables wound on a winch that looked like a big yo-yo on the rear of the cat.

The operator set the choker on a log, winched it to the cat, then clomped the cat on to the next log. He repeated the whole process until a half dozen or so logs were tethered to the crawler. Then the cat headed on down the hill, with the blade still biting, tracks pivoting and screeching, with the butts of logs rooting a deep furrow straight down the fall line. Anywhere else, that furrow would be called a ditch, an instant watercourse inviting erosion. Gallacher's cameras churned, recording scene after scene that was not supposed to exist. Over the weeks we had been on the story, we had asked industry people repeatedly about just such practices, and they denied them, said that forestry was about enlightenment and stewardship and respect for the land. Yet we stood on the hillside that day and saw only a torturing of the land, soil sent downhill by the blade of a cat or churned and left naked and helpless before the force of the first rushing thunderstorm.

Gallacher and I spent a couple of hours documenting the destruction, and then we left Gold Creek. I began to understand skid trails and the significance of having this scene multiplied across Montana. Those trails scoured by the cats are repeated every hundred feet or so on almost any steep clearcut. At that rate, a section of land, a square mile, can accommodate more than two hundred such trails, each a quarter mile long. Champion International owns about thirteen hundred sections of land in western Montana. So does Plum Creek, and each plans to have cut all of it by the close of the century. The force of this multiplication can only sink in after you have stood in an old skid trail, as I once did, where rain and cats had eroded it four feet deep, laying bare a cross section, first of detritus, topsoil and organic decay, then of volcanic ash, then of glacial gravels. The wealth of the eons squandered in a few passes of the cat that wrings a few more pennies from an eight-foot two-by-four.

Each of these thousands of skid trails runs down hill, as does water, first to ephemeral streams and freshets, then to creeks, then to rivers like the Big Blackfoot. There what we knew as topsoil becomes silt, and fish die in the translation. Silt gathers in the spaces between rocks where trout deposit eggs, smothering both the eggs and young trout. In a few years, once vibrant trout streams become sterile ditches, broken by the work of carrying topsoil away. This river was the setting of Norman Maclean's "A River Runs Through It," a story of family, place, and fly fishing. Now trout fishermen bypass the Big Blackfoot because the fish are mostly gone. Biologists suspect sediment from logging is at fault.

In interest not of logging but of cheaper logging methods, we squander our soil, trees, and fish. An honest man—corporate or otherwise—probably would admit this is unwise, and he did, or at least Jim Runyon did. Runyon was the forester and corporate official Champion designated to speak to me about all of this when I told the company I was investigating its timber practices. I still don't know what to make of that choice, because Runyon was, at least by the standards of other corporate sources I have dealt with in my years as a reporter, unusually forthcoming. Runyon was also a man in a bind, and he talked first about that on a Friday afternoon in May a few weeks after we had visited Gold Creek. Gallacher and I met him at his office in Champion's Montana headquarters, a simple wood building in sight of the Bonner mill. Runyon is a tall, friendly man whose kind eyes are set off by a handlebar mustache and wire-rimmed glasses. In jeans and Oxford-cloth shirt, he settled comfortably into conversation.

Still, he toed the company line, spouted the logic that drives timber corporations, and internally at least, that logic is compelling. There are variations on this theme within the industry, but basically the argument boils to this: to survive in an international economy one must offer the world market competitively priced goods, be they two-by-fours, computers or hamburgers. Accepting that condition, though, indentures a local industry to the harsh rule of international prices, a particularly onerous burden for Montana's timber industry. Western Montana is a place only barely hospitable to trees. Compared to more sodden regions such as the Pacific Northwest, our trees grow slowly, sparsely, out of the way at the end of long, steep, and expensive gravel roads. Our timber-producing lands are on the economic margins, which shows up in higher costs of production for our wood products. If the area is to hold its own in international markets, then it must find ways to shave costs. To a large degree, the industry does this by mining logs in the cheapest way it can.

If the great invisible hand of the market were as rational and unifying as some articles of economic faith would hold, it seems a practical solution to the problem in our forests would emerge. That is, society simply would not produce lumber from marginal lands, but would satisfy demand from more productive places. In a free-market utopia, the marginal lands would be priced out of the game, at least until a shortage pushed the price to the point that loggers might operate on marginal lands both responsibly and profitably.

Unfortuantely, this is not a unified system. Champion and Plum Creek, with their considerable investment in Montana, are not about to bow out in deference to Weyerhaeuser or Boise Cascade and those latter companies' more productive land further west. Champion and Plum Creek's executives will tighten the screws, scrimp, save, and cut corners to stay in the game. They must compensate for the handicap dealt by the rocky, steep, and dry land. It is what the corporate system expects from them. They will say what industry has always said: they do this to avoid closing mills, to save thousands of jobs and prevent the collapse of the local economy. They may not be doing it for precisely that reason but it doesn't matter, because if the mills did close, that economic upheaval surely would happen. There is pain in this, because guys like Runyon are right: guys like Rausch have a choice between a disagreeable job and no job at all.

What does this invisible hand of the international market care for our trees, or for men like Runyon and Rausch for that matter, and who should know that better than they do? In Runyon I encountered resignation. The preconceptions that took me into this story and what I saw happening on corporate lands led me to expect predatory archetypes of capitalist mythology overseeing all of this. To a degree I did find such people. Just as often, though, I found people who behaved like prey, who saw themselves on a lower niche in the food chain of laissez faire capitalism.

Runyon weighed in on the latter side early in our interviews when he acknowledged without equivocation (unlike Plum Creek's Bill Parson) that, yes, Champion had decided for strictly competitive reasons to cut all of its own lands in the matter of a decade or so, far faster than those trees could grow back. He said the strategy was based on a cold calculation that once its trees were gone, Champion could use its economic and political clout to pry loose a supply of logs from public, especially Forest Service, lands. The reason was simple: Champion did not have to pay for logs from its own lands, so could use that economic advantage to survive depressed prices for wood products that emerged in the early eighties.

"It was one of those situations where somebody had to make a decision," Runyon said. "We may not want to do it, but if we want to be here and survive tomorrow, we may have to."

Other sources within Champion said that the early part of the decade also brought a fundamental shift in attitudes within the corporation. That is, foresters who once were encouraged to think long term, to consider the next generation of trees, were told vociferously (in one case, in literally a table-pounding session) that "long-term" was defined as next year. It was the shift in philosophy that sicced the cats on the hills.

That table-pounding punctuated a shift that went down hard for foresters, who are by inclination and training more geared to growing trees than cultivating P&L statements. In our first interview, that conflict rang in Runyon, both a forester and executive. I had brought the slides that Gallacher had shot of the cats. After about a half hour of relatively innocuous questions, I dropped the photographs on his desk. He looked at them for a long time then joked with me: "Tell me these were shot on Plum Creek's land."

Then he stared at one shot of a cat's blade gouging the slope and he began talking like a forester, assessing with an expert's eye the performance of the cats:

> If he [the cat operator] is dropping his blade, then the impact is going to be a whole hell of a lot greater. There's no question about that. It looks

> like he's basically creating just a huge erosion path
> through there . . . . He's taking the soil and it's go-
> ing to end up down in Gold Creek. You're right. I
> don't make any bones about that. And as a forester
> that ticks me off, because as a forester I was taught
> a long time ago that everybody says I am managing
> a stand of timber, but that's not the true resource.
> The resource is what the trees are growing in and
> that's the soil . . . . If he just rips the hell out of the
> topsoil, then I have lost a fair portion of my
> acreage [but] to say that does not occur would be
> lying . . . . I'm not going to lie. It happens.

Here is a statement that cuts straight across the grain of the
public myth that shelters the industry. We suffer the cutting of
trees here only to the extent that the industry is able to grow
new trees. The industry spares no expense in public-relations
gimmickry designed to convince us of this, even to convince us
that they grow better trees once they have replaced nature's ill-
designed forests with tree factories of their own devising.

The cats crush that myth. Runyon, a man whose life to a
certain extent had to be organized around it, still acknowledged damage. I pressed him no further because I believed I
was dealing with an honest man. It is a failing of my reporter's instincts, but my experience has made me wary of aggressive questioning of honest people. Likely, they will
answer my questions, and sometimes I don't want them to do
that. If they give me answers, then I will print them, and they
will be chewed to bits by their own colleagues. The system
goes hard on blunt people.

Runyon, nonetheless, wanted to talk more, and he called
me a few days after the interview. We met for a beer in downtown Missoula. He started talking about some of the problems of overseeing operations on eight hundred fifty thousand
acres, how things can go bad despite the best intentions.

Sloppy contractors. Loggers disrespectful of the land. But then he went further, disclosing without prompting something I had not known nor suspected.

When Champion bought its lands from Anaconda, it did not know what it was getting. Much of the land was too harsh for trees, even by the rocky standards of Montana. Much of the land flowed over dry, south-facing slopes in areas more suited to grasses and sagebrush. Some trees did grow on these places, but by only owing to quirks: come a few wet years in a row, a cold summer of the sort that happens once in a century, and a few seedlings catch. A forest then rises by the strength of its own moisture-holding shade, an up-by-the-bootstraps operation. Champion was clearcutting such places, Runyon told me. The accident of nature was being reversed.

Methods and foresters' artifice did not matter in such dry places. Likely, once cut, those sites would never see trees again, at least in the sense humans understand "never." In the more limited corporate definition of "never," the cutting was even more final. Runyon had looked at those lands and decided that, once they were cut, the corporation should let them revert to counties for back taxes. In the internal logic of the corporate world, those lands would be all used up. A process that began with catastrophe—the explosion of a mountain seven thousand years ago—ends with a cheaper pine stud on a lumberyard's pile. I wonder about the way we use the word "catastrophe," why the volcano and the fires that gave life are listed as catastrophe while the taking of the life by the cats is not.

I wonder how we will guard our trees and, more important, the forest. Are humans to be trusted with this task? We have given our trees to the corporations, and they are so harried by the predations of international markets that the next quarter is their only hurdle. We have entrusted the forests to foresters who believe that they are visionary in ensuring the next generation of trees. They generally mean to protect the existence of "rotation-age" trees which live about sixty years, just long enough to be cut. McLeod and his buddies, the copper kings, once cut Gold Creek's trees, and now Jim Runyon and Kevin Rausch cut them again. We ask no more than this continuity of a single century.

But what is a hundred years in the lifespan of a tree? Who will stand guard for five centuries? The land that holds Mahala Jane Primm's grave is an island, a few acres among one hundred square miles of hard-cut Champion land. Standing on this island, one can see the surrounding storm; one can see nearby slopes, pounded and slit by the cats. All around this island of peace, history is crashing in, not held off by but drawn by these few hundred ancient tall trees. They would make fine saw logs, no doubt.

Still, the stubbornness of Frank Parker kept the timber thieves at bay, and a miracle happened. Parker held out until 1938, when his life was laid down in this soil. Mahala Primm and her husband took over in 1939. In their late years, there was domestic trouble of sorts and Charles Primm left for a bit, a separation that lasted until he died. Some friends lured the old woman away from the homestead for an afternoon, and Charlie Primm came home when the friends slipped his body into the soil. Then Mrs. Primm held on alone in her four-room log house fifteen miles from the nearest highway up a rutted gravel road. Fifteen miles from the nearest house and the nearest power line. Something about these trees kept her here, even until that last winter when her fingers all froze. That was 1977, the year that she too would lie down in this soil. She was a year shy of eighty.

Her grave rests among these trees that have lived for several centuries, that have lived so long they have created their own separate peace with the earth, a peace not just of trees, but of the forest, the magic symbiosis of animal, plant, and soil worked and worked through the centuries until it is a web of relationships that we humans and our histories and our economics, our science and our journalism understand not at all. Or maybe some of us do, because here the symbiosis included humans. The Primm's homestead was not a sanctuary for trees; it was a working farm that sustained humans. There are stumps here; the Primms did cut trees. Yet after nearly one hundred years of use, their place feels like sanctuary as unspoiled as any national park. Here is land ruled somehow by an understanding of the centuries, the centuries of life that corporate timetables do not consider.

Now who will guard those centuries? Plum Creek and Champion in the next quarter? Runyon in the next rotation? You and I who can understand only our time, a puny lifetime when measured against that of a tree? Or what about Mrs. Primm, who was buried where she lived and died, among those ancient trees. Something about her will lives on and tries to keep the saws and the cats at bay, tries to protect those few ancient trees that in turn protect her grave and the soil that holds it. Is her will stubborn enough to prevail two centuries more when some of these ancient trees will die, lie down in this same spot to rot, enrich and enliven the soil that will feed the next generation? Can she rule here as long as our nation has lasted?

Can she rule five hundred years after that to bring the next generation of ancient trees? Or seven thousand years beyond that to allow this earth to bring back the soil ripped from the hills around Gold Creek? A few years after Mahala Primm died, her heirs sold this land to Champion. I fear this island of peace will come to know the whine of the saws and the screeching, ripping cleats of the cats. I may be wrong about this, though.

Champion bought the land in 1977, and it has sat unlogged ever since. Then, late in 1990, when the company began taking a public beating for its timber practices, it made an announcement that may indicate some evolutionary progress toward a land ethic. Champion said it will preserve the Primm property as a special research area never to be logged. At the same time, the company declined any legally binding measures that would ensure that intention survives through further shifts in corporate thought. It is still too early to relax the guard Mrs. Primm maintained throughout her life, but as of this writing, there remain trees worth guarding in a pocket of Montana's mountains.

CHAPTER FOUR

ON A BRIGHT MORNING in spring we flew low into the sun. Before this, we had learned of the clearcutting of Montana from what people had told us. We had seen it in the numbers tucked into computers and heard it hashed in the debate between conservationists and loggers. We had walked among the trees and the clearcuts. There was, though, still another perspective. Michael Gallacher and I flew that day to find that nothing could tell this story as starkly as the view from a Cessna 210.

The four-seater rose from Missoula's airport, wafted east along the Interstate, over the steam and the sea of logs piled outside of Champion's mill at Bonner. It cleared the hills that pinch the low end of the canyon of the Big Blackfoot River, then pointed upstream above the web of logged streambeds and ridges. I sat in back, turtling my arms and neck inside my Gore-Tex parka, full winter battle dress in spring, but necessary against the altitude's chill. I hate flying with photographers. Gallacher sat up front on the right side, his seat belt loose enough to let him stick his head and lens through the window he'd propped open, the same hole that dumped a considerable measure of the Arctic jet stream into my face. Now and again Gallacher asked the pilot to dip his right wing low and curl the plane into a tense bank, then he shot through the open window the plane's slant had tipped straight below him. His motor drives churned, winding frames of the images of clearcuts, skid trails and stumps that sheeted the panorama. Flying low and slow, away from the worn paths of airliners, we saw what few ever do. All the way to the rising sun stretched evidence of an ill-concealed sin.

Then we wound away from the Blackfoot and into the See-ley-Swan Valley, away from Champion's land, over Plum Creek's. There were differences, such as fewer skid trails rutting steep slopes. Plum Creek's land is flatter, and the company is not as aggressive as Champion in sending the cats onto the ridges. The sweep, though, is the same: a valley floor woven of clearcuts, a patchwork that cloaks the shoulders of the limestone peaks rimming each horizon.

Then came a bass thump on the skin of the Cessna just beside my ear. I lurched against my seatbelt. Gallacher laughed. The thump was his doing; he had leaned too far out the window and lost his Denver Bronco cap to the wind that slammed it harmlessly against the fuselage. We flew on, suspended by the hypnotic drone of the prop, by the surrealism of this vantage. I watched the story I must understand roll slowly by me in a stream of images, as if I were gathering them on one of Gallacher's film spools. The details began to jell, and then I realized I had seen this view before, not this color, but this shape. From the air, Montana has come to resemble Forest Service maps that are color-coded to show land ownership. I was looking straight at the checkerboard, drawn not on a map, not an abstraction, but real on the land below. The ebb and flow, the contours, the winding of river valleys, the random roundness all lay violated. The land now bears like jail bars the surveyors' linear notions, sheer lines, section by square-mile section. Where Champion and Plum Creek owned land, they had cut land, leaving straight boundaries visible from miles above. Clear-cut clearcuts, straight edges slammed against wilderness, the roll of hills rammed into square holes. Montana's face had been cut up and sold.

DISCUSSIONS OF ENVIRONMENTAL DILEMMAS INVOKE THE RULE of the tragedy of the commons: when all members of a community are given common access to a resource, even overwhelming sentiment to protect the resource will not ensure protection. One person can always gain by exploiting the resource and will use his right of access to do so. A commonly held grazing area at the center of a village generally is cited as the best example of the tragedy of the commons. The will of the vast majority of farmers to conserve the grass will not ensure its protection as long as some are willing to overgraze the common plot. And some people always are. A better example still of this tragedy is the handling of Montana's timberlands, especially those clearly common, public lands overseen by the U. S. Forest Service.

Gallacher and I flew that day away from the Seeley-Swan Valley and over the Mission Mountains, across the broad Flathead River Valley and then to the lower Clark Fork River Valley just west of Missoula. The Cessna banked up a tributary of the Clark Fork called Fish Creek. The Forest Service manages much of the land there. It is our land, logging administered in our name. The commons. Yet in some places our land is cut every bit as hard as the corporate land, clearcuts swatched hard and square. There is a reason for this: the Forest Service, not the corporations, pioneered the use of the clearcut as a "legitimate" logging practice in the West. The experiment began not far from Fish Creek on the Flathead National Forest.

People think of the Forest Service as a sort of collection of overgrown Boy Scouts, rangers sheltering trees, flowers, and Bambi from errant campfires and carelessly flicked butts, the prototypical Smokey Bears. In the West, though, where most of its domain lies, it is difficult to consider the Forest Service anything but a branch of the timber industry. Nationally, the agency superintends enough land to fill the states of California, Washington and Oregon, about two hundred thirty million acres. Each year it sends to mills enough timber to build about 1.2 million houses. It has not always been so. When the Forest Reserve Act of 1891 established what are now the national forests, they were considered just that: reserves. They were what was left of once far larger public holdings sold or traded to the timber companies. It was assumed that the companies would conduct the timber business and the government would protect forests and the watersheds the forests sheltered. Even Gifford Pinchot, father of the Forest Service and the social engineer modern-day environmentalists love to excoriate, did not envision the national forests as forage for mills.

In the agency's early going the forests flourished in obscurity. Then came the post-war years of prosperity, procreation and the Eisenhower-esque notion that the business of America was business. The timber industry's appetite, whetted by the post-war housing boom, grew beyond private lands' ability to feed it. And then a fortuitous wind storm leveled a city-sized patch of Engelmann spruce on the Flathead Forest, a storm that lives on in aftershock. The Forest Service's then-novel approach to those downed trees in the mid-fifties was to log them all, every stick. This produced huge amounts of timber and popularized what we now know as a clearcut. The technique had been used before, but mostly when the goal was to strip the land for farms or grazing. Until that windstorm, clearcutting hadn't been accepted as sound tree-farming within the Forest Service. Nonetheless, the agency genuflected immediately to the new-found religion. It soon began clearcutting even before windstorms prepared the way.

It was the Forest Service, not corporations, that in courts and Congress fought and won the battle of the clearcut. By the mid-sixties, use of the technique was epidemic and alarming, especially to long-time residents of mountain valleys that were coming to resemble bad haircuts on punk rockers. The controversy erupted in the Bitterroot Valley, which lies just south of Missoula. Senator Lee Metcalf, a progressive Democrat from Montana, appointed a special commission to investigate. The commission, headed by Arnold Bolle, then dean of the University of Montana's School of Forestry, concluded that the Forest Service's clearcuts were excessive, were ravaging the public's lands. The report made international headlines when Metcalf released it in 1970. Senator Frank Church of Idaho used it to draft guidelines limiting the use of clearcuts, though the Nixon Administration refused to use them. Congress subsequently drafted them into law, but the law still was weak enough to allow the Forest Service, by then addicted to clearcuts, to carry on.

Then the battle moved to the courts and to the East. The National Resources Defense Council, an independent environmental group, learned in the mid-seventies about a forgotten provision of an 1897 law. The legislation required the agency to mark every tree it cut and allowed the killing of only mature trees, a provision ignored in clearcutting. The NRDC then located particularly egregious clearcuts in the Monongahela National Forest in West Virginia, sued, and won. The court banned clearcuts. The Forest Service went back to Congress and won a new law allowing clearcuts, but with some restrictions, chiefly that, with few exceptions, clearcuts can be no larger than forty acres.

From the Cessna as we flew that day, we could see all of this history written on the hills of Fish Creek, giant clearcuts, some twenty years old, some with no new trees, splotched across the hills. Even more striking than the clearcuts, however, was a landscape sewn together by roads. Not that I didn't know about the roads. Gallacher and I had spent months lurching over more than three hundred miles of these winding washboards. But in these travels, we always saw but a road at a time, never the whole labyrinth displayed. We had been the rats in the maze, but now the Cessna was showing us the maze. The hills stood terraced with roads like the decks of a pyramid, roads that wind around the hills like a peel spiraling round an apple, cut and cored. Cut and fill operations these roads are called, cats and graders biting gouges straight out of the side of a hill and laying the gravel in swaths across draws. Then the log trucks roll on this new bed of grit, sand and dust.

The U.S. Forest Service is the world's biggest roadbuilding agency. Nationally, it has to date built 340,000 miles of roads, a realm eight times as long as the nation's interstate highway system. The agency plans to stretch the web by 100,000 miles during the next fifty years. Virtually everyone I talked to on all sides of the logging controversy agrees that the greatest damage from the timber harvest is the logging roads it engenders. ·

Roads kill fish, there can be no doubt about it. These roads are raw, steep faces of loose sand and gravel ground to powder by the tread of logging trucks. They all lead down hill, and in the northern Rockies there is a stream at the bottom of every hill. Rain washes the grit to the creeks, sealing the spaces between rocks, normally holes where fish spawn. One study in Idaho showed that logging roads devastated the fish of the Salmon River and were still hurting them a full twenty years after all logging in the vicinity had ceased.

Roads kill elk. Montana now has a severely shortened elk season largely because of logging roads and clearcuts. The logging has deprived the animals of places to hide, but the roads compound the difficulties. "Sportsmen" simply drive into once remote places and blast at the elk through pickup windows. Roads provide avenues for noxious weeds to work their way into once pristine habitat. Spotted knapweed, an exotic species that is a plague in the valleys of western Montana, has worked its way up logging roads like gangrene up an infected leg. Roads convert the thousands of square miles they cover from habitat to hard-beaten sand flat. Roads are a blight on our once-wild lands, and yet from the air, they appear a dominant part of the landscape now, as permanent and implacable as the hills they wrap. They are the mark of the tragedy of this commons.

That clearcut in the early fifties in Montana was an event of no small significance, not only because it was ancestor to these clearcuts and roads we see today. It marked a significant and now entrenched cultural change within the Forest Service. It rode the crest of the post-war baby boom, an era of material demand, of new families and therefore a need for new houses and therefore more lumber. It also rode a Zeitgeist that convinced us we had a right to dig, bulldoze and saw every square foot of our nation and everyone else's nation to derive for Americans the goods that were their god-given right. Arnold Bolle characterized the transformation in the Forest Service as a shift from Stetson hats to hard hats.

A conservation ethic dominated the Stetson-hat era. It is hard to imagine now, but conservation was the main vein that fed the Forest Service until the fifties increased pressure to log. Then the bureaucratic mindset shifted. Word sifted down from the top that there would be rewards—literally rewards: plaques, citations, raises, and promotions—for those people who could get the wood out. There was considerable resistance to this from the agency's old guard, but the rewards of exploitation operated on the commons. All it took was a few young turks to see the advantages to their careers in breaking the old rules. The men who pioneered that clearcut became heroes in the agency. There was a cultural change that persists today, the culture of the timber beast. To this day, Forest Service managers are evaluated and promoted on their ability to get the cut out. If a manager is not willing to do so, to exploit the land for his own gain, then there is always someone right behind him who will.

Granted, there are checks to the system, especially on Forest Service land. Two decades of environmentalism and court battles have brought pressure to bear, have dragged the agency to a grudging admission of its responsibility to protect the integrity of our forests. Two decades of a growing environmental awareness have produced a new breed of sensitive biologists filtering into Forest Service employ. These people form a counterculture nibbling away at the preserve of the timber beast. Still, the timber beasts sit atop the flow charts, and the beasts cut trees, no matter what the cost. In Alaska, the *New York Times* remarked, you can buy a several-hundred-year-old public tree cheaper than you can buy a hamburger. In Idaho, lodgepole pine in the Targhee National Forest sell to timber companies for less than a dollar a tree.

I know a man named Bob Wolf, a talkative old gent, a forester by training and a bureaucrat by habit of his career. He spent a working life watching the Forest Service from Congress, overseeing research on issues relevant to public lands. As a staffer, he drafted much of the nation's key public lands laws because he once believed it was in the nation's best interests to allow the Forest Service to cut trees. In retirement, though, Wolf says he came to an old-man's understanding of trees, that they are more than board feet. Now he spends his quiet years plodding through the same stacks of numbers that filled his career. He uses these numbers to goad the Forest Service. Wolf has a simple bottom line: he says he will believe the Forest Service has mended its destructive habits when it stops lying about how much money it loses selling public trees.

For the record, the Forest Service itself admits to losing money on timber sales in ninety one of the nation's one hundred fifty-six national forests. That is, on those forests, the costs of building roads and overseeing cutting are greater than the market value of the timber, the price the timber companies bid and pay to the Treasury. According to the agency, however, sales on the remainder of the forests, largely those holding the old growth of the Pacific Northwest, more than cover the losses. They produce a "profit" on the order of $400 million a year.

The Forest Service's numbers, however, are cooked. For instance, administrative costs are hidden. Effects of logging on wildlife are considered "positive," because the timber beasts argue that logging improves habitat. Those "benefits" are entered on the profit side of the ledger. Capital costs are amortized at odd rates. The costs of roads in the Chugach National Forest in Alaska were once spread over twelve hundred years, thereby levying infinitesimal costs against each year's proceeds. Wolf used accepted accounting methods to run an eleven-year cash-flow analysis of the Forest Service timber program. He says the agency's claim of a profit is simply false, that the timber program racked up a deficit of $3.3 billion during that eleven-year period, $5.6 billion with interest.

The beneficiaries of this largesse are the timber mills, and why should they turn it down? Why should a single mill owner buck the system only to go out of business and allow another mill to exploit the niche? This is the commons and willing exploiters wait on line. Why should a timber company take pains in its cutting to adopt more careful methods when those methods are more costly and therefore render his lumber less competitive? Why should a Forest Service inspector take pains to enforce more than minimum standards on a logger when that will lead to fewer board feet being cut? The greater the cut, the more money the Forest Service gets to spend on roads and bulldozers and desks—growth. When more is cut, the Forest Service gets a bigger budget, the ultimate goal of all bureaucracies. Deficits accrue to the Treasury; growth accrues to the bureaucracy.

Still, this public tragedy of the commons is but half the plague. It spreads to private land by what is best called the tragedy of the common stock.

NORTHERN CALIFORNIA—CHARDONNAY AND HOT TUBS notwithstanding—is a land of loggers. The state's redwood region cuts about 4.5 billion board feet of softwood lumber a year, more than triple Montana's total cut. More than half of this comes from lands owned by corporations, including a new and significant player, Maxxam. This corporation's new lands were formerly those of Pacific Lumber Company, privately and locally held in Scotia, California. Pacific believed itself to be a part of the community and so practiced what amounted to heresy in the carnivorous business milieu of the eighties. Pacific owned about seventy-five percent of California's still-standing redwoods outside of national forests. The company carefully budgeted the cutting of its trees. It practiced sustained-yield logging, taking even fewer trees than natural growth would replace. It practiced selective logging, taking a tree here and a tree there in a pattern that was unnoticeable. Pacific believed in supporting the human community as well as the natural community. Employees were guaranteed such benefits as free college tuition for their kids. Then Pacific Lumber made a mistake. It entered the corporate commons with a public issue of stock.

Pacific was debtless, held substantial assets, and earned solid profits. In short, it was a target for the takeovers rife in the mid-eighties. Rising to the bait were some of the all-stars of that rapacious decade. Charles E. Hurwitz, a Texas oil and real-estate tycoon and takeover artist, began buying huge blocks of stock. Pacific resisted the obvious takeover attempt but finally fell in 1985. Hurwitz produced $795 million in notes financed by Drexel Burnham Lambert, the junk-bond firm that would later figure in the matter of the federal investigation of Ivan Boesky and the conviction of Michael Milken on six federal felonies.

Hurwitz had won, but only by amassing huge debts. He began dealing with that debt by liquidating his "assets," in this case those carefully husbanded forests. Suddenly, Pacific's lands, long an object of community pride, were indistinguishable from other corporate lands in the region. Hurwitz clearcut Scotia.

There is a curious aspect to the accelerated harvest of corporate lands: it was not unique to Champion, Plum Creek, or even to Montana. Virtually every tree-holding corporation in the Pacific Northwest started mowing simultaneously. The coincidence suggests collusion, but it probably reflects instead a common understanding of reality. In a time of asset-strippers, timber was an asset the stock market had greatly undervalued, because timber is a precariously held asset. Fires could burn it. Bugs could infest and kill it. A swell of environmentalism could produce laws preventing it from being cut. Over the long haul (and trees require nothing if not long-term commitment) it was nowhere near as safe as money in the bank. The corporate solution, then, was to convert trees to money in the bank.

A manager no longer had the option to manage for sustained yield, for the long term. A company could not husband that "asset" in perpetual harmony with the needs of the forest. Instead, the manager could either cut the asset himself or face a takeover by someone willing to do so. The corporations cut.

THE CORPORATION I WORKED FOR WHEN I WAS LEARNING ALL of this was Lee Enterprises, which, at least ostensibly, deals not in timber. Lee owns newspapers, or "media properties," as they have come to be called. Headquartered in Davenport, Iowa, the corporate realm largely consists of a string of daily newspapers in Midwestern farm towns. As Anaconda Copper Company was retreating from Montana, it sold its trees to Champion and its newspapers in Billings, Helena, Butte, and Missoula to Lee. That gave Lee four of the state's seven daily papers, considerable corporate clout. Then, however, Lee was sensitive to rampant anti-corporate sentiment in Montana and so guarded the independence of the four papers. The corporation was not so much interested in wielding the power of the press as it was in making money. This benign neglect was the *Missoulian's* greatest asset.

Missoula is, by western standards, a progressive, liberal, and well-educated town. Most of this is owing to the presence of the University of Montana, but also to a certain quirky liberalism that has grown from Montana's penchant for progressive farm and labor politics. It is the sort of town

Lee is least equipped to understand. The corporation owns nineteen newspapers, centered mostly in the cornbelt. Its corporate sensibilities evolved in places like Carbondale, Illinois; Muscatine Iowa; Bismarck, North Dakota; and Winona, Minnesota. Within this circle of "properties," the *Missoulian* had the reputation as the corporation's eccentric paper, staffed by the hippies and tree huggers, the avant garde. Still, the *Missoulian* made money. The corporate powers did not understand Missoula, and so they let the paper operate in more or less a state of anarchy.

This made the paper a fine place to work, a reporter's paper. Salaries were far better than average, which attracted a staff of veterans. There was money for ambitious projects, investigations, and quality. Editors were demanding. News was treated seriously. In the mid-eighties, when I joined the staff, it did not seem possible these halcyon days would die. Catastrophe, however, was already in full rage elsewhere in the business.

Although coverage was absent in America's newspapers, a development of some significance to our republic unwound quietly during the last decade: American newspapers, especially small American newspapers, went to hell. An era ended. It had blossomed in the wake of the Vietnam War and journalism's dethroning of a sitting president. The altruists who honed their anger during the civil-rights and anti-war struggles found there was a place for further struggle in newspapers. In the mid-seventies, when I began reporting, the publisher of my first paper posted in his office a slogan from the old *Chicago Times*: that the duty of a newspaper was "to print the news and raise hell." He was not a hell-raiser, nor was his paper, but he let me raise some, an act that provoked my love affair with reporting. I joined a craft that seriously considered it its responsibility to foment democracy. I was privileged to do work that mattered.

My first publisher, though, was a dinosaur; different forces were ascending. Businesspeople learned that ownership of a newspaper is a license to print money. Two trends bore this lesson. First, competition among papers collapsed, leaving most cities with only one paper. Where competition remained, it became subject to "joint operating agreements." Those allowed seemingly competitive papers to combine certain operations, to agree to divide the business pie. Thus, by 1990, ninety-eight percent of all American cities were either newspaper monopolies or subject to joint operating agreements.

These monopolies were more profitable, yielding the industry's second major trend. Corporations flocked to the candy dish to grab the remaining family-owned or independent papers by the fistful. In 1960, corporate chains controlled only about thirty-two percent of American newspapers. By 1986, chains held seventy percent of the papers; in 1990 they held more than eighty percent. Further, the big chains got bigger. By 1986, the nation's ten largest chains controlled forty-three percent of the total circulation in the nation. In 1968, the ten biggest had controlled less than thirty percent.

Much has been made of the threat held in this great concentration of power. Newspapers are regarded as the nation's shapers of opinion. This is, I believe, an over-rated threat. There are now too many competing sources of information. Newspapers are being sold to an increasingly apathetic public that ignores a publisher's views, assuming any publisher has views extending beyond his P&L statement. In fact, there is no clear evidence that the chains have exploited their monopolies to set the nation's political agendas, nor is there evidence that they ever sought to do so. There is clear evidence, however, that they sought monopolies to boost profits, a strategy that has worked.

By 1990, newspaper people were calling their craft "the most profitable legal business in America." Writing in the *Washington Journalism Review,* former *Wall Street Journal* reporter Jonathan Kwitney noted that newspaper publishers were expecting pretax profits ranging from twenty to forty percent of net revenues, while some of the more predatory chains were turning nets at fifty-nine percent of revenues. At the same time, the likes of General Motors, Exxon, and IBM were all showing pretax profits far less than half that large.

Newspaper profits created a feeding frenzy. First, a few corporations found that a combination of exploiting monopolies, aggressive marketing, and cutting newsroom budgets could create licentious profits. Stockholders pressured less exploitive corporations to adopt similar strategies. The *St. Petersburg Times* is a fine example.

Legendary in the business for its commitment to quality journalism, the paper was the legacy of Nelson Poynter, an old-style editor. He died, and Andrew Barnes, an editor in the Poynter tradition, took over. Times, however, had changed. The paper's list of stockholders came to include one Robert M. Bass, a corporate raider of some renown and player in such high-stakes games as Donald Trump's Plaza Hotel, Bell & Howell, Macmillan, Disney, Time-Warner, and the savings-and-loan debacle. Bass played off a festering family feud in 1988 to win forty percent of the newspaper's stock, then immediately mounted a court battle to win a majority ownership. Bass's aim was to stop the *Times* from such activities as plowing 83 percent of its profits into the Poynter Institute, a center promoting journalistic excellence, and spending enough money on newsroom budgets to win the paper three Pulitzers. In the end, the faithful at the *Times* were able to force Bass to, in the words of Forbes magazine, "buzz off." In doing so, though, they paid what amounted to greenmail, $75 million for the stock Bass had bought for $28 million. That removes almost $50 million from the budget for journalism. Life at the *St. Petersburg Times* will change, the cost of a Pyrrhic victory. The paper has entered the commons.

THE CORPORATE REINTERPRETATION OF THE FIRST AMENDment caused at least casual notice among some members of the general public. Their newspapers became fluffier, more geared to "lifestyles," shorter stories (known as "quick reads"), more pictures, more color, more breathy, bold-faced accountings of the comings and goings of celebrities. These changes, though, fell harder on those of us in the business. In 1988 Doug Underwood, a former reporter for the *Seattle Times,* Gannett News Service, and then on the faculty of the University of Washington in Seattle, wrote in the *Columbia Journalism Review:*

> It's not surprising that, as corporations have extended their hold on U.S. newspapers, the editors of those newspapers have begun to behave more and more like managers of any corporate entity. It's understandable, too, that in an age enthralled by the arcana of scientific business management—and at a time when the percentage of the population reading newspapers has declined—newspaper executives have reshaped their newspapers, in the name of better marketing, more efficient management, and improving the bottom line.
>
> So maybe we shouldn't be worried as the pressures grow on newspapers to treat their readership as a market—to use the words of the business consultants who have proliferated throughout the industry—and the news as a product to appeal to that market.
>
> Well, after spending more than a dozen years as a reporter with the *Seattle Times* and the Gannett Company, I'm plenty worried. And, after interviewing more than fifty reporters and editors around the country, I find a lot of others who believe that profit pressures and the corporate ethic are fundamentally transforming—and not necessarily for the better—the nature of the newspaper business as generations of reporters and editors have known it.
>
> In fact, many of the people I talked to say they feel increasingly unwelcome in a business that once was a haven for the independent, irreverent, creative

spirits who have traditionally given newspapers their personalities.

Underwood's seminal piece was called "When MBAs Rule the Newsroom," a title that chilled me when I read it early in 1988. It was only eighteen months before then that Phil Blake had taken over as publisher of the *Missoulian*. Blake is standard-issue MBA, slick, efficient, properly grayed, pinstriped, and brief-cased. The *Missoulian* had been more or less profitable under its previous publisher, Tom Brown. It was not, however, wringing from the community the profits that newspapers were making elsewhere. As profits and takeovers mounted at other papers, this sort of sustained-yield newspapering was no longer acceptable. The commons demanded a maximizing of profits at the expense of everything else. The days of Lee's benign neglect ended, replaced by a drive to make the *Missoulian* look and behave like a chain newspaper. Explicitly, this meant that we were to be forced into the hole carved by *USA Today*, Gannett's experiment in puff journalism that had by then spread like syphilis. *USA Today's* mix of glitzy art, how-to stories, trite headlines, and quick reads had become the accepted get-rich-quick formula of the business. The *Missoulian* was in fact a hold-out, fighting explicit pressures from the chain to ape *USA Today*. More than most chains, Lee wished to follow that trend. Underwood revealed this in his *Columbia Journalism Review* story:

> ". . . We're trying to put out a newspaper for a whole new generation of newspaper scanners out there who expect to develop a conversational knowledge of what's in the paper based only on reading the headlines," says Dan Hayes (editor of Lee's flagship, the *Quad City Times*). "The reality is that newspapers are no longer at the center of people's lives. Other things are."
>
> If this sounds like a version of *USA Today* applied to the local level, it is. Stuart Schwartz, the

marketing director for Lee Enterprises . . . says,
"What Gannett is doing with its national newspa-
per, Lee is doing with its local newspaper. I think
Lee is leading the industry right now."

Shortly after Blake took charge of the *Missoulian,*
Schwartz and a gaggle of spear carriers jetted into Missoula
to found a new order. Management ordered the staff to re-
design the *Missoulian* to resemble *USA Today.* The staff
fended off the incursion as it always had in the past: listened,
bitched, made a few jokes about Schwartz, then ignored him
to conduct business as usual. This time, though, things would
change. On a Friday evening the week the new orders went
down, the corporate entourage swooped into the *Missoulian.*
That night's news editor, Bill Keshlear, had already designed
and sent Page One to paste-up. It was, of course, a page that
violated corporate policy. Management ordered it pulled
back. In earlier days Brad Hurd, the paper's editor, might
have intervened on the side of integrity and independence of
the paper, but apparently he had lost his stomach for taking
such stands. Keshlear redesigned the page.

There was, by coincidence, a gathering of reporters at my
house that night. Reporters generally finish their jobs before
editors do, and so we were several cases of beer into a newsroom party by the time Keshlear arrived to tell his story. It
felt as if there had been a death in our family; our party
turned into a wake.

It was in this environment, at the end of this decade of the
corporate takeover of newspapering, that I wrote my stories
about the timber industry in Montana. Despite growing evidence to the contrary, I still believed it possible to print them.
The spirit of the *Missoulian* had not died yet. Hurd had still
given me the time I needed to work on a lengthy investigation. By May of 1988, I had been working the story full time
for three months. Reporters at most small papers don't have
such luxuries, and I held that as clear measure that there still
lurked in the *Missoulian* a commitment to printing news and

raising hell. I believed that commitment would allow me to report what the timber corporations had done.

Just a few weeks before the writing, I read Underwood's story in the *Columbia Journalism Review*. It had hit hard. It had revealed that what I had assumed was a private war waged by a few malcontents and iconoclasts at Missoula, Montana was in fact infecting the whole of my craft. There was no refuge. It should have chilled me even more than it did. There was a quote in Underwood's piece that did not stand in particular relief then, but rings when I read it now. Commenting on the spectre of increasing corporate control of newspapers, a former *Dallas Morning News* reporter told Underwood: "Do you see a corporation that's in the business of making money going out and investigating other corporations? I don't."

IN MAY OF 1988, I FINISHED MY INVESTIGATION OF THE COR-porate decisions to log western Montana. The story was this: both Plum Creek and Champion had abandoned sustained-yield forestry and were logging their lands as rapidly as possible. The practice was doubly damaging in that it would, within less than a decade, create a widespread shortage of logs that would undermine the long-term health of the cor-porations' own industry. Further, the corporations had logged with wanton disregard of accepted environmental standards. This harmed the natural systems of the forests and so in the long run undermined those lands' abilities to re-grow trees. I spent about a week sequestered in my house writing a series of seven stories that built this case. They were stories I believed in more than any others I had written. I knew them to be true and I believed they held vast implica-tions for the health of the community. Still, I had no idea how the *Missoulian* would treat them. On the one hand, the paper was still enough of its old feisty self to print them. On the other, the new corporate order was exerting itself. Blake's boosterism was laying an increasingly heavy hand on the newsroom. My stories ran hard against the boosterism that was beginning to pervade the place.

I wish I could say my stories then triggered the great clash of values that the setup promised. Crusty editors shrieking at overzealous reporters and such, a play of passions straight from *The Front Page.* Certainly I am capable of, or more accurately thrive on, such excess. I always wore a pugnacious interoffice style that had dragged me into more than one screaming match with editors concerning matters of newspapering. Such passion was, I believe, a part of the charm of the business, the best indicator that what we did was goddamned important and certainly worth screaming about. But the corporatization of newspapering has bled it of its passion. Its heart is out, surviving now solely with a calculating head, a head for business.

When I submitted the stories, they provoked silence. City Editor Brian Howell, a friend of mine who at the time was having his own struggle with the new order, had the first look at the series. Howell had worked heavily with me throughout the investigation, so he knew what was coming. I submitted the stories to him late in May, expecting to spend a week editing, then watching them go to print within another week or so. Instead I faced week on week of silence. During this period, I often saw Howell and Hurd cloistered in the latter's office, which was a glass-walled fishbowl on the edge of a gymnasium-sized newsroom. They'd pore over computer printouts of my stories, pencils poised, faces flat. Clearly something was up. Both Gallacher and I fretted. We worked internal office sources for information. No help. Howell avoided me for nearly a month. Finally, though, he confirmed what I suspected: the stories were in trouble. I wanted to be angry with Howell then, to hold him as a sellout and a part of the conspiracy to eviscerate the *Missoulian.* For a bit, that's what I did. I also had built enough respect for him to believe he was still an honest journalist and enough of a realist to temper my excesses. After a bit of cooling off, I began to listen to his advice, counsel that I now believe saved the project.

His chief objection to the stories was that I had written them unconventionally, more a magazine style than a straight-forward just-the-facts-ma'am accounting. He was right about that. The information was unusual, and so I believed it deserved unusual treatment. He and Hurd were especially uncomfortable with the fact that much of the work depended heavily on my own direct observation and research, rather than on quoting "experts" who had broken the ground ahead of us. To a certain extent that was unavoidable, because Gallacher and I had worked unbroken ground. Logging on private lands in Montana is unwatched. There were no independent experts to quote. To a certain extent, however, we could work around the latter problem in a rewriting of the story, and it was clear that I had to rewrite if the stories were to have the slightest chance of seeing print.

Howell offered me a solid assessment of the internal politics of the *Missoulian,* and he was finally blunt about that. He wanted the series published, but he suspected that Hurd and Blake were looking for excuses to kill it. Howell and I then agreed on a strategy of surrendering to their objections in the minor battles to win the war. I rewrote the stories. At that point, Howell signed on to the project, in fact became the advocate, the pilot who could navigate the series through the corporate politics that I did not understand. It was a crucial development because by that point, I could not plead my own case. Management regarded me as, in Howell's words, a "bad boy."

There was some history to this. Just before Blake came to the *Missoulian* in the fall of 1986, I had written a column critical of the publisher of the *Billings Gazette,* another Lee paper. This won me no friends in high places. Then, shortly after Blake came, I learned that he had become a member of a cabal of executives lobbying the legislature for some changes in Montana's tax structure that would favor business. When a state legislator openly charged that the editorial page of the *Missoulian* was a tool of this lobbying effort, I wrote a story quoting the legislator and Blake's denial of his charge. Blake spiked the story on the grounds that publishing it was not in the *Missoulian's* best interests.

These transgressions of mine created a climate wherein it was best that I avoid any direct contact with Blake or Hurd where my series was concerned, a tactic explicitly agreed on by Howell and me. Howell argued. I kept my mouth shut, as it turned out, for nearly four months. Both of us quietly agreed we would would resign in protest if the series did not see print. During this whole period I had no idea what was happening to work that had been the most intense effort of my career. I paced the newsroom and wondered as the closed-door meetings in Hurd's fishbowl ground on. There were no screaming matches, no healthy ventings of belief and passion like the ones I had always used to distill some truth through the imperfect art of journalism. There were only a few tense looks, and then memos on memos on memos.

Through Howell, I made compromises. For instance, he said Blake objected to the fact that I had caught Plum Creek's Bill Parson first denying, then confirming, that the company had abandoned sustained yield. Blake demanded that I not report Parson's original denial, only his subsequent confirmation. I believed the reversal was relevant to my story for what it revealed about Parson's credibility. Grudgingly, though, we acceded to Blake and sanitized my account of Parson's transgression. Blake objected to the number of Gallacher's photos we chose to accompany the stories. We believed the photos to be a crucial element of the series in that they were the smoking guns, direct evidence that the timber company's cats were violating accepted forestry practices. Howell told me Blake said our use of so many photos made it look as if we were making too much of the whole business. We ran fewer photos. Howell reported that Blake said during one meeting with him, "What's the big deal if a little bit of gravel gets into the stream?"

In the end, though, none of the changes gutted the series. By altering emphasis and rearranging some of the more inflammatory detail, we were still left with a vital piece of work. Finally, as summer wound down, we had an investigative series that Blake and Hurd could no longer refuse to run. Still, publication dates kept being pushed back without explanation. Still they stalled. There was a reason.

One of the most upsetting facets of the story was its prediction of an imminent timber shortage, hard times ahead for the local mills. Newspaper executives' bottom lines depend on a solid local economy, or at least a perception of prosperity, so this is the sort of news they least like to print. It was particularly upsetting in this case because I was alone in this conclusion; it contradicted more optimistic assessments by industry analysts. In fact, the Missoula Economic Development Corporation, a business-booster group, was about to issue a report predicting mills in our area faced nothing but clear sailing well into the next century. Blake knew this because Blake sat on the board of the Missoula Economic Development Corporation. Hurd ordered my series held until that report was issued, in effect undermining his own paper's work in favor of the boosters' work.

I considered the commission's report a curious development in that I knew some of the people preparing it. Some of them understood that the future was not at all rosy for loggers. Nonetheless, the group issued its preliminary report, and it did indeed contradict what I had concluded independently. I called a source for an explanation only to find that the commission's politics were similar to those at the *Missoulian*. Spin control had been exercised. The development group was, after all, in the business of attracting business to western Montana. Issuing reports of a gloomy economic future was counterproductive.

In the report, the group based its prediction of an adequate timber supply on one very large assumption, that the Forest

Service would meet its sales targets for public timber. The problem is that the Forest Service has never met those targets, averaging, for a variety of reasons including environmental constraints and litigation, about 30 percent below target. Further, federal law commands the Forest Service to compensate for damage from overcutting on private lands by cutting less on public lands. That meant the supply of public logs would soon shrink even further.

My source said those who prepared the report were fully aware of this. When the commission came up with its rosy prediction, it created an internal howl that led to the report being withdrawn. A compromise was reached. The report would be reissued, spelling out the fact that it was based on the dubious assumption that the Forest Service targets would be met. That satisfied the dissidents' consciences. At the same time, the development corporation knew full well that press and public typically read such a report's bottom line but fail to question assumptions. With spin thusly under control, the report went public.

After several contentious meetings, Hurd ordered me to write a story about the report. I objected, but compromise was the rule of the day. I wrote the story, still pointing out the assumption that seriously undermined its conclusion. With that out of the way, the last objection to running my series evaporated. In October, nearly half a year after I wrote it, an eternity in the daily newspaper business, my series went to print. About the same time, ads appeared in the paper announcing that the *Missoulian* had joined forces with Champion and Plum Creek to sponsor a major basketball tournament at the University of Montana.

FOR THE FAITHFUL, NEWSPAPERING CAN STILL HOLD MOMENTS of high romance. One can best feel this passion in the somnolent hours of early morning, when a foot on a silent city street can feel the rumble of a printing press somewhere below. Through the years, it was my habit to follow those rumblings to their source on nights when a story that mattered was being measured out in ink. Late at night, after the editors had left and the newsroom went dark, I scrambled into the building between trucks at the loading dock, then down a long flight of steel steps toward a monstrous basement cavern that roared and shook as if someone had imprisoned a locomotive inside. Air brakes. Screaming electric motors. Chains. Sledgehammers. Rolls as heavy as Volkswagens shooting rivers of paper toward inked drums. From the center of all this, feeding a conveyor headed upstairs, flowed an endless chain of daily editions, ten thousand every hour. Each copy held word of garage sales, a horoscope, school lunch menus, girdle ads, obituaries, and a bit of truth as best we could find it.

We call this dried ink information. I think now about what lies behind this owning of presses, about how we have come to regard our information. This once simple act of writing, an act of love, now seems less lovely. I think of information as it was once contained in wood. The floors in my house were laid in 1913. Scuffed, beat, and worn, once even suffering a shag carpet, they survive. They still read like a newspaper, or more accurately, volumes of history. The grain of wood is information, and sometimes I sit on the floor trying to read this organic record of a community, an honest recording of conditions in Montana during the last century, a grain that is really an array of growth rings strung with the caprice and randomness of nature. A good year here, a bad one there, drought, rain and fire. In our time, though, such wood is becoming scarce, and so we are losing the patient record it holds. It has became fashionable to compensate for this loss with hardboard, ground-up bits of small trees pressed and glued together by machines. Then, on the resulting four-foot-by-eight-foot sheet, a press prints information, or rather someone's imitation of information, a simulated grain of wood, a false accounting of centuries that never occurred on trees that never grew, a work of fiction. Yet it matters little that this imprint is false, without integrity. It sells, and in our time that has become the measure of value.

I have always hated hardboard for its manifest dishonesty, and yet it was a hatred that came to haunt as I felt the crush of the weight of change in my own business. Newspapers are becoming less and less organic accounts of the ebb and flow of community. They are becoming more information that sells, false grain pressed on ground-up trees.

When I began writing the timber series, I always assumed my objectivity, that I wrote as an outsider to the timber industry. I was a newspaperman, not a timber beast. And yet, after watching a press run one night, I walked into an adjacent room full of rolls of newsprint, each about four feet in diameter, tipped on end and stacked on top of each other to form a wall of continuous tubes reaching to the twenty-foot ceilings. Stacked as they were in that dim light filtering from the ceiling, they resembled nothing if not trunks of great trees, and so they were, timber to be felled and fed to a corporation's presses.

CHAPTER FIVE

THERE WERE TENSIONS in my life after publication
of the timber series. My belief in the validity and integrity of
my craft was slipping. In my private life there was trouble
relevant to the troubles everywhere. It was not a unique dif-
ficulty, but uniquely felt. During the week of the publication
of the timber series, a judge downtown granted a decree in
my name; a marriage of seventeen years ended. This divorce
mined the same sad vein as did the crumbling of my career.
It too came before love had ended, but after I realized love
must end. Certainly these troubles qualified me for a seeking
of the sublime. I suppose I was seeking, but I do not remember doing so. I recall being confused and disoriented, yes,
but no seeking of the ineffable, for me, the forest. I have
only the sense that the forest, or what I will come to call the
dreamtime, sought me.

The timber stories issued to silence from the industry. My
sources said this was because the stories' revelations were an
open secret within the corporations, information they never
denied but were not anxious to disclose. Publicly the industry
suffered the shot quietly; privately it stewed. Meanwhile, environmentalists, wildlife officials, Forest Service staffers, and

independent analysts praised the series, a source of satisfaction for me. At least on the surface, my work was going well. There would be no explosion for another year, but already I had felt the rumbling just beneath my feet. I went about my business, but what I had learned about nature and about my newspaper had changed me. It was not to be business as usual.

Hints from the forest were steering me in a new direction. Deeper questions began to nag. During the investigation of the timber series I met people on both sides of the issue who spoke sincerely. Some were outraged and pained at the sight of a clearcut, with the knowledge that Montana's hills had been slicked off and sold. Others, just as sincerely, I believe, could look on this carnage and feel no pain. I wondered about the visceral line that divided people this way. I wanted to know what drew that line. Why do some of us feel in our guts assaults on nature while others do not? Was the difference genetic? In hearts and minds? In training? Was it formed by corporate logos on paychecks?

After the publication of the series, this question kept welling in the back of my head. Yet I knew as a reporter I could not openly ask it. It lay outside a newspaper's visible spectrum. Newspapers—another way of saying the public debate as it has evolved—have a well-defined universe, a horizon or boundary to their reality. Within that bound, we are permitted to conduct a vigorous debate. My years in the midst of it, however, taught me to distrust this limited debate's ability to resolve important questions. Yet the troubles of our time and the urgent questions they have generated demand resolution. An expansion of the universe, a redrawing of its boundaries to permit the asking of broader questions, is now necessary. It is time to consider a separate reality.

This seems altogether clear to me now, but was not at the time I write about. Then, I only suspected that broader questions mattered somehow, but I had no plan to redefine reality. I suspected that the troubles epidemic in our people, in me, were altogether wrapped in this, that there was something about our exploitive relationship with nature that explained the bankruptcy of our relationships with each other. This, however, remained an issue a newspaper reporter could not explore and so the broader questions, at least in my work, went unasked, at least unasked by me. I now have only the impression that those questions asked and began to answer themselves, that a series of seemingly unrelated stories I would cover for the next year eventually brought me to reckoning with the possibility of a separate reality.

THE ABORIGINES OF AUSTRALIA SPEAK OF A DREAMTIME. IT IS not so much a time, however, as it is a place, or in our Western way of understanding, a concept. We live in a perceived reality, a finite universe defined by laws that we understand. Our lives are defined by cause and effect, predictability and control. Dreamtime, however, is no less real, no less subject to law. On the contrary, it is said to be more real than real. It is, without our knowing it, the source of the laws that govern what we call the real world. The portion of the universe we understand is simply a shadow of the sublime forces of the dreamtime. It is not a concept unique to Australia. Native American people understand the limits of reality in much the same sense, hence the importance of shamans. The shamans are the seers, the people who move back and forth between the two universes, between the mysterious and the mundane.

Poet Wendell Berry explains this another way when he advocates a deference to mystery, or as some would choose to define it, religion. He argues that there is a boundary between that which we understand and that which we do not. Everything on the other side of the boundary is mystery. Over time, our knowledge expands, and the boundary moves, or rather, we gain a firmer sense of where the boundary really lies. Still, there is always a portion of the universe that does not yield to our tools for understanding, our ways of reckoning. Our brains and senses are subsets of nature, creations of nature, and so are limited. We are not the whole and so cannot comprehend the whole.

Our failing, the single failing that allows the abuse of nature, is refusing to admit the existence of the territory on the other side of the boundary. We deny the existence of mystery. From this failing rises hubris, the notion that we are in control, the false extrapolation that says because we understand some things, we understand all things that matter, that we can account for all that is relevant. Our undoing is our failure to acknowledge that some of what is vital and relevant lies beyond our grasp in the dreamtime.

I understand now in raising this that I bring this conversation perilously close to what friends of mine would call "woo-woo," a sort of contrived spirituality that seems especially to issue from the western coastal regions of our nation. I wish to say that this notion of dreamtime is not necessarily freighted in activities such as crystal gazing, the wearing of war paint, whooping and pounding on drums. I am not at all sure that even these are exercises spiritual, because I no longer know what "spiritual" means. In our time, the term "spirituality" has been so promiscuously abused that it has been bled of its meaning. In my own mind, the term's common border with words like "intellectual" and "physical" has been so often breached as to void the distinction. And so by invoking the dreamtime, I do not mean to force the attitude of prayer. I wish to suggest the necessity of taking into account the unknowable. I only wish to say that we have spent these years debating matters concerning trees without accounting for forests.

I AM BOTHERED ESPECIALLY BY THE SMUGNESS OF FORESTERS. They claim they know all about trees, and I suppose they do. Their training assures they do. They spend years scratching bark, sniffing needles, measuring limbs and reading each others' theses. Foresters know the hundreds of species on sight, not to mention on smell or even by feel of the grain. They know of water tables, climax, density, rotation age, release, of canopies and uneven age. This information they use to grow trees. Sometimes they do it well, or at least well by the standards they have set. Trees are planted. In sixty years or so, childhood for most species, they grow to something that can be fed to a mill, more and more a chip mill, a large enough stem to be ground and glued into something resembling a board.

Foresters have done this long enough—in this country, about a hundred years—that they now feel confident to overrule the hand that nature has exercised through the millenia. They do this by planting trees, a matter of pride for the profession. During recent years, it has become fashionable for the timber corporations and professional associations of foresters to commission advertisements in environmental magazines. These generally feature a photo of a tender hand on a vigorous young sapling. A copy block announces the millions of times this nurturing has occurred in recent months, that the companies and foresters are concerned with the future and the future is their planted trees. We are advised that wood is America's renewable resource and that every day is Earth Day for a forester.

There is a certain cynicism behind these pronouncements. Many of these planted trees will not grow, and the companies know it. Champion officials, for instance, admit they have cut exceedingly dry slopes in Montana, areas that once grew trees by the accident of a few wet years. Once stripped of the snow-holding canopy of existing trees, these baked slopes will remain deforested. Even on friendlier sites, accidents happen. A drought follows the time of planting and so all the young trees die. By the time planters are able to revisit the site, brush takes hold and chokes out the effort. Or somebody makes a wrong choice about strain and species. Somebody misreads the wildly varying conditions of mountain slopes. Then the variables such as aspect, slope, and rainfall wipe out a plantation. Timber companies, however, do not commission advertisements about these failures.

A Forest Service forester once took me on a tour of his mistakes, clearcuts that would not regrow trees after decades. I remember one such site, but it was covered with green, a dense stand of twenty-year-old Engelmann spruce. This he called a failure. To a trained eye it was. The ends of the spruce trees were twisted like arthritic fingers owing to an attack of spruce budworm. The forester's mistake had been to replant that clearcut with trees grown from seed taken several drainages away. Through the years, local populations of trees evolve immunities to local bugs and disease, immunities the young trees from just a few miles away did not have. For a mistake made twenty years ago, this stand of spruce would die. This is how foresters live and learn. Their decisions must stand tests of long time.

Yet as I have struggled to understand the term "forest," this issue of failure of a plantation has become less bothersome. I am more worried now that the foresters will succeed, that the trees will grow as they have designed them to grow.

Tree-planting operations are spooky sights. They consist of a half dozen or so young people, young because the pay is low and the physical demands extreme. Each planter carries a bag of hundreds of half-foot-tall saplings and tool called a "hoe-dad." The planters walk the draws and ridges all day slamming the spud and planting the trees. They are paid a few cents for each tree planted.

Their workplace is a Dresden of sorts, a ground zero stripped of trees. They work on black ground, first logged, then bulldozed, then burned. A forest once lived here, three to eight species of conifer, species the loggers cut for mills and species that just plain grow. Together with the trees grew dozens of species of brush and shrubs that hug the forest floor. Ceanothus, ninebark and serviceberry, shrubs that display nipped twigs, evidence of the browsing of deer, elk and moose. A forest holds bear grass and its breast-like flowers, paint-brush, lillies, peas of endless variety, huckleberries, orchids. A forest holds the dead, downed, rotten logs fringed with fun-gus, standing snags, dead-topped trees, bug trees and wood-pecker trees, hollowed grounds that are, in Montana, home to more than fifty species of mammals, eighty-five of birds, other vertebrates beyond number. A forest holds cool, wet air.

On logged ground, all of this is gone. The plants have been cut and scraped, the animals driven away by loss of food and shelter. The limbs, dead logs and snags have been piled and burned. Now man will create a forest in his own image. In the bags of the planters are one or two species of trees, a forester's sole provisions for replacement of all this variety of life. The tree planters march on straight parallel and imagi-nary lines eight feet apart. Every eight feet, they jam a sapling into a spudded hole, now points on a grid. A plantation. A monoculture different from a Midwestern cornfield only in species, slope and spacing. I have always reacted viscerally to this act of taming, of supplanting the wild with the staid and predictable, but let us be fair. What do my tastes matter in this? Is this not anthropocentric of me, that I only recoil at the compulsiveness of a tree plantation because in my life I have fought against similar treatment of people? Let us be hard-nosed, because, after all, this is commerce. Our nation needs the trees. Might this grid not be the best way to grow them?

The dreamtime, the aborigines say, is not unreal, it is hy-per-real. It is the place where, without our knowing it, the rules for reality are set. Foresters, at least the predominant species, are not keepers of forests. They are tree technicians. And so the forest is the dreamtime of the trees.

THERE IS AN IDEA FLOWING IN ZEN BUDDHISM, AN IDEA OF A sort of triggering tone or encounter. If the mind is open to enlightenment, there will come from an unexpected corner a single simple sound to change the world. Maybe the rattle of a leaf or the twist of a flower in a hand. Or more often, a single song from a single bird. This lone epiphanous trill rotates one's vision a quarter turn to spill the line of sight beyond the plane that held everything heretofore seen. One animal that can trigger a new vision of the forest is the red-backed vole, or more precisely, the vole taken together with its colleagues the Douglas fir and the truffle.

As deforestation in Europe ground to its final phases, forestry, at least as we know it in this country, began. Well-meaning people set aside plots of trees in France and Germany to manage them for continuous growth, recreation, and lumber. A striking feature of these was their park-like countenance. Foresters virtually vacuumed the forest floor, removing dead and decadent trees, downed limbs, anything that threatened the contrived order. These forests are the pattern that schooled generations in this country. Today the European plantations, places like the Black Forest, are dying. Trees will not grow. Generally this is held to be the fault of external environmental factors such as acid rain. The trees are counted casualties of industrialization. However, Chris Maser, former government forester in the Pacific Northwest and a seminal figure in an evolving school of holistic forestry, says these trees are dying because of traditional forestry.

To be sure, acid rain is hastening the demise, but pollution causes even more trouble because those trees are already seriously weakened. There is a range of causes for this, most brought on by our ignorance of the vital relationships that drive a natural forest. The relationship between trees and fungus describes this problem best. In a healthy forest, fungi attach themselves to the roots of conifers to consummate a tree's bargain with the soil. The fungi, aside from collecting nitrogen-fixing bacteria, send out a fungal root system. This system collects minerals, breaks them down to a form usable by the trees, and then sends them on to the trees. In trade for this service, the fungi derive sugars from the trees. Symbiosis.

To begin with, fungi depend greatly on the organic nutrients and the moisture-holding capacity of humus, of decay.

This humus derives from the undoing of dead trees, that same decadence that those compulsive European foresters had removed. In his research of native forests in the Northwest, Maser commonly found thirty to forty species of these fungi attached to the roots of a healthy conifer. In Germany, unhealthy Norway spruce roots in a managed forest were the home of only three to five species of fungi.

This support network, however, is far more than a matter of trees rising from their own dead. We think of these fungi as mushrooms, but a mushroom is only what biologists call the "fruiting body," the apple compared to the whole apple tree. Fungi have a large subterranean network that makes up the entire fungus. The function of the fruiting body, aside from providing food for a range of creatures, including humans, is to propagate the fungus. It contains the spores. Certain of these fungi fruit above ground, and so their spores can be scattered by the winds. Certain others, the fungi we call truffles, fruit completely beneath the earth. Humans find truffles in Europe with trained pigs, but this is of no satisfaction whatever to the trees whose roots are dependent on the spores' dispersal. The trees need propagation of truffles, and evolution has seen to it that the trees themselves will indirectly ensure the spreading of the spores.

Pigs can find truffles because when they are ripe, they send off a smell very similar to that of an amorous boar. A variety of other creatures have evolved a sensitivity to this same smell. In the Pacific Northwest, the trees shelter the mouselike red-backed vole. It finds, excavates and eats truffles. Then in its wanderings and its diggings, the vole drops the remains of digested truffle, including still-living truffle spores. Each pellet of vole feces contains about 300,000 viable fungus spores. In its digging, a vole often encounters the roots of young conifers. Vole droppings contact roots and the next generation of fungi attaches itself to the life of a tree.

The red-backed vole, though, is a creature of the forest. It lives in trees. It is among the first species to leave or die after loggers clearcut. Banished, too, is the flying squirrel, which also eats fungus and spreads spores. Clearcuts banish the deer mouse, which performs the same service. Clearcuts banish the black-tailed deer; a single pellet of its feces contains about 10 million viable fungus spores. Foresters have long noticed that clearcuts seem to regrow trees best around their edges, adjacent to an uncut forest. In this tale of feces and spore lies a hint as to why this happens. Forest denizens like voles, mice, and black-tailed deer wander from their uncut sanctuary and into the clearcut, slowly spreading the web of life that supports them. Yet the tale of the life of a vole is but one unimaginably small part of the life of a forest. Vole-truffle-tree is but one small circle in a long spiral of interwoven cycles. The life of virtually every bird, bug, and beast is ringed by vital cycles. There are thousands.

The pileated woodpecker, for instance, controls carpenter ants and spruce budworm, but at the same time chops great square holes in standing dead trees. These eventually become nests for a whole string of mammals and smaller birds, which in turn serve the forest. Pileated woodpeckers are becoming rarer because they nest only in standing dead trees larger than thirty inches in diameter. Forestry being what it is, such standing trees are becoming rare. Meanwhile, the insects the woodpeckers control become more numerous.

This web of interdependence eventually begins to blur the lines we would draw between species. Maser points out that each individual termite is three creatures in one. Termites have strong jaws and so can chew wood. The problem is, they can't digest it. A protozoan lives in the gut of the termite and handles this task. The protozoan, however, requires nitrogen, often not present in sufficient quantities in decaying wood. The termite's gut is also home to a nitrogen-fixing bacteria that completes the effort. If the triad, a natural chord, did not exist, then the termite could not exist, and in the forest the termite is not a pest. It begins the process of breaking down wood, effecting the decay that eventually will come to rest on the forest floor to sustain coming generations. Termites decompose wood to send it through the nutrient cycle again.

To human eyes, these termites and the range of tree-attacking insects and diseases are pests. We would prefer that pests be eradicated. Yet in the forest, what we call disease is simply a vital part of the web. It infuses the system with death and decay. The web runs on decay.

We can see this in an individual tree. A century-old fir is considered to be at rotation age, a point when its adolescent growth spurt slows. If foresters had their way, and mostly they do, this is the age at which all forests would be cut. The effect on the forest community is analogous to the effect of removing all people older than fifteen from the human community. Foresters insist that allowing trees to live longer is waste because they begin to grow more slowly, become "decadent." Truly they do, and that is precisely the point. A Douglas fir can live eight hundred years, yet "live" is a somewhat arbitrary term. Even in its youth it begins to die, infusing the system with the attending benefits, almost as we humans absorb the wisdom of an old person who has accepted death.

Maybe the tree's top goes first. At, say, two hundred years, a chunk breaks off and insects and fungus attack the wound. Rot sets in and a hawk takes up residence in the tree-top cavity. Over the centuries the rot spreads, maybe around an old fire scar at the base. Fungus spreads. Carpenter ants work a piece of the tree. Woodpeckers go to work, followed by nesting squirrels and bluebirds. A storm wipes out some limbs, maybe half the tree, letting some additional sunlight filter to the forest floor. This encourages a young tree already rooted in the fungus and detritus building at the tree's base. Maybe a century later, this long process crosses a line we draw and we say the tree is dead. Yet is it? By now it is honeycombed with new life, a standing city spread clear through with mammals, birds, insects, and plants, all woven to each other. Now the tree is a snag, and as such will stand for another two hundred years, if it escapes a firewood-cutter's saw or escapes being toppled by a logger's dozer. Loggers fell snags not for milling, but because they might fall and harm bulldozers or loggers.

But our snag is lucky and the centuries pass. Now, thoroughly perforated by the life it holds, the snag does fall. The cavities opened by the insects begin to do their work. The fiber digested by the termites becomes a sponge and now the downed log is a reservoir storing spring's snowmelt, rationing it to the forest's next generation. Sometimes these downed trees are called "nurse logs," in that some species of saplings are wholly dependent on their care. These logs nurse a row of fresh green saplings like a sow with a litter of pigs.

Still centuries later, as many as four hundred years after its death, a tree finally crumbles until we can see it no more. But the forest sees it. Now it has completed the millenium it took to round the cycle to rest in the soil again. One thousand years that the concept of rotation age has reduced to one hundred, the tragedy of seeing not forests, but trees. In all of this life and death of a tree, in its planting, its growing, its aging, its bit-by-bit dying, its standing and its falling, this tree drives and in turn is driven by uncountable whirls and loops of cycles of lives. The loops, the uncountable permutations and combinations of living cells weave together the concept we call forest. Is this not mystery?

Granted, it can be argued that this mystery's unraveling really lies in the question of numbers of relationships. Perhaps the forest is but the measure of our ignorance, not unknowable, but in our blindness urged by the rush to cut trees, as yet unknown.

To an extent this is true. In the time our people have been practicing forestry, we have accounted for only a very few of the thousands of relationships. The insights that have fomented this new holistic school of forestry emerged only during the past decade, pushed by a handful of scientists working mostly in the Northwest. Its founders are people like Maser and, within the Forest Service, Jerry Franklin and Hal Salwasser. The latter are responsible for much of the work leading to the drive to protect the spotted owl, and even more important, have pushed the enlightened system of forestry within the Forest Service. Working under the flag of the conservation biology movement, these scientists' first axiom is that one cannot consider the health of an individual species without accounting for the health of the system, the web of relationships on which it depends. This is not just about wildlife such as spotted owls. It is about our ability to grow trees, and so the loggers' future rests in this web as well. We know almost nothing about the web, but then we have only begun to look. Those relationships we have examined are understandable; they lie within the ken of our science. All of this may be knowable, after all, only a matter begging further investigation.

Perhaps. There is, however, in the sheer power of numbers, an element of the unknowable. We can comprehend a star or two or a thousand but when the known quantities multiply by known distance they yield numbers beyond our imagination. Like the stars, the forest is driven by vast numbers of relationships, each playing off the other in a bedeviling matrix expanded in geometric progression. How can we possibly hope to enumerate, let alone explain, all of these combinations? Yet this is more than an issue of sheer numbers. Like the stars, the forest holds a mystery beyond our simple inability to count. The forest seems driven by ideas the human brain refuses to hold. It seems haunted by puckish spirits that appear obvious on the periphery of vision but invisible when we stare them straight on. It seems haunted by somber spirits that our wills to survive refuse to allow us to contemplate fully.

A FEW MONTHS AFTER PUBLICATION OF MY TIMBER SERIES, I was on assignment. A group of people went for a walk in a forest. Marcia Hogan, who is a Forest Service press aide, a friend, and a woman who loves forests, had arranged the trip. Besides Marcia and me, the group included Dick Hutto, an ornithologist from the University of Montana; Mike Hillis, a Forest Service biologist; and Rosalind Yanishevsky, a Ph.D. who specializes in old-growth forests. We walked up a closed logging road just southwest of Missoula to a stretch of the Lolo National Forest that has never been logged. The Forest Service has set this land aside to meet legal requirements for preserving pieces of old-growth among its logged lands.

As we approached this land, Hillis spotted a goshawk cruising at treetop, a good sign. In summer, a goshawk is a raptor of the deep forest nesting in broken-topped trees and feeding on the scurrying city of small mammals and birds that the forest floor sustains. Hutto heard mountain chickadees and returned their call, showing us how to distinguish it from the cleaner call of the black-capped chickadee. We found some ancient snags, mostly Western larch, fully colonized with smaller bits of life. Piles of shredded wood fiber, the borings of ants and beetles, heaped from the duff around the snag's base. Huge chips obviously thrown by the great bill of a pileated woodpecker dusted the earth.

The *éminences grises* of this place were the larch trees, maybe one hundred or so feet tall, four hundred or so years old, some three hugs around at the base. These trees were dying, that was clear, dying of nothing more profound than their vast age. Most showed only a few live limbs toward the top, broken tops, dead slabs, ants already at work. Within the next century or so, this generation will pass in the obscure luxury of falling on the ground that raised them. For more than the next two centuries, longer than our nation has been in its making, they will languish in their dying, rot, and fall, lives laid down in the soil.

Beneath the canopy of larch stood middle-aged subalpine fir, a shade-tolerant tree, a coming generation asserting itself. And beneath the fir an impossibly dense tangle of fallen limbs, shade-killed young trees, and long-dead elder trees composed the mix of shelter and detritus that is the forest floor. All of this, said Hillis, brought us smack against the dilemmas of understanding an old-growth forest. Left to its own devices, this forest would very soon become something other than an old-growth forest. The problem is that in the most recent century the forest had not been left to its own devices. What we were examining, primeval as it appeared, was artifice.

There was something wrong with this picture. It was not the dying larch. They were behaving as they should. Their time for dying had come. The problem was those vibrant young fir. Were this forest on wetter slopes, there would be no problem. In moist areas of the region, fir belong, but not, at least not as a climax species, in most of Montana, a dry place and so a place ruled by fire. Fir grow here, but in a natural forest typically flushed by fire every twenty years or so, not many survive to old age. Since the Ice Age, forests in this spot raised up huge larch and ponderosa pine, dry-land trees resistant to more frequent fire. It is only in this century that man's work of fire suppression has changed the face of the land. This brings us to a strange irony: environmentalists tend to blame loggers alone for the demise of the old-growth forests. To be sure, loggers have done their share, but so has fire suppression. Our urge to preserve trees has precluded the future of old-growth in the Northern Rockies.

How do we deal with this? Or to move to the question implicit in this, the more basic question: what do we mean by "old-growth?" How do we define the forest?

Clearly, we mean something other than simply big, old trees. In California's Yosemite Valley, a place that annually draws 3.5 million people, there stand some marvelous trees, ancient ponderosa and Jeffrey pine, cedar and sequoia. The visitors, though, have stripped every scrap of vegetation from beneath them, pounding the forest floor to a flat-packed pan of bare dirt, a biological desert. The trees are but a vestige of the forest that has died. This place is no forest.

Conversely, can there be a forest without trees? In some places in Montana, the normal course of events may well include a catastrophic fire that wipes out all living trees, leaving bare ground. Some species of trees have evolved to reseed quickly into burned areas; in fact they depend on the fire to prepare their way. Is not this bare ground simply another stage of the forest? When windstorms flatten mile-square swaths of trees, also a common occurrence here, is this not a forest, too?

Hillis's role in the Forest Service is to ponder and deal with these questions as best he can, and in doing so he raises another paradox. Given the fact that this stretch of forest has been reshaped by fire suppression, what choice does he have but to attempt to re-create a natural forest with a manifestly unnatural act? What would be wrong with commissioning loggers to gently cut and snake the young fir from beneath the ancient canopy, clearing the way for the next generation of larch? Hutto and Yanishevsky are skeptical about this plan. Both experts on old-growth, they still maintain we know too little about old-growth to re-create it with such a heavy hand. Further, given that virtually all ecosystems of the Northern Rockies have suffered fire suppression, where will we find our pattern, our "control," for this bold experiment? Where is the example to tell us what to do?

Or are we misguided in seeking forests untouched by humans? Fire suppression is a white man's invention, but native people meddled too. On this, the record is clear. Since the Ice Age, people have been in the region, and for at least that long, they have set fires. From northern Alberta to Texas and Southern California, native people historically used fire to reshape their environment, to kill trees deliberately. What then, given this long history of control, is "natural?" What is an old-growth forest?

Yanishevsky took a stab at a definition, one that has hung with me solidly ever since. We can go into a forest that shows a certain set of characteristics and begin piecing its history. A windstorm that dropped a tree that made a small clearing, sunlight and a path to the sky for smaller trees. Fire in this century. Disease in the last. Outbreaks of bugs. Woodpeckers.

Voles. Hawks. A tree falling in the forest unheard. The pattern that emerges from all of this is no pattern at all. We look long and hard for the regularity in the forest, but this is looking in the wrong direction. We ought to be looking for irregularity, for caprice, for lack of rhyme or reason. This seems more to describe nature. Yanishevsky defines an old-growth forest as the living record of a series of random events.

And now I think of the tree planters trying to re-create a forest by spudding holes on an eight-foot-by-eight-foot grid. How can this mathematically monotonous pattern of a single species, a monoculture, re-create the randomness and diversity of the forest? In fact, how can anything the human brain devises be random?

This is a live question being chewed upon by some of the nation's mightiest computers. In 1988 a polyculture of disciplines—statisticians, computer scientists, probability theorists, physicists, philosophers, and even psychologists—assembled in Columbus, Ohio to ponder a stubborn problem. True, some progress was reported at the session, but still the problem remained. No one had yet devised a system to overcome the human brain's bias for pattern. That is, no human or human machine has heretofore generated a truly random series. The naked human brain falls well short in the task of duplicating caprice. When asked to rattle off a random series of, say, ones and zeros, respondents invariably do so in a readily discernible pattern. Most of the time order emerges first as an obvious pattern designed to avoid a pattern.

Machines are not much better. No roulette wheel is perfectly round. Patterns emerge from machines after a very large number of repetitions. Such unimaginably long strings are not at all irrelevant where the forest is concerned, where endless generations must sustain life through millenia.

The closest we now come to randomness is through computers. For years, programs that purportedly generate random numbers have been available, but all have been flawed. All produce discernible patterns. Now larger computers programmed with ever more arcane and elegant mathematical formulas are being employed. Science is finally approaching an act that nature performs effortlessly.

We are not given to pondering the implications of our acts through the millenia, although our acts clearly do have such implications. We are not equipped to see why randomness matters to the forest. We can only see in this issue of randomness something that the forest is and that we are not. True, there are patterns in a forest that we can understand, the small circles and cycles. These are our teachers, the root of our empirical science, but these are only pieces. The threads are orderly, but the weave is random. It is the whole, the dreamtime, that we cannot understand. Further, we do not understand the implications of this ignorance, and so we wrongly believe there are no implications. This is our greatest mistake.

CHAPTER SIX

THE DROUGHT SUMMER of 1988 shot forest fires through the Rocky Mountains. To contest this advance of nature, the federal government spent roughly half a billion dollars. Much of this money trickled into Missoula, a four-state region's command center in the war on fire. Aside from a twinge of prosperity, this role infused Missoula with a mission and the brisk stride that goes with it. People here tend to work military metaphors to death, but a raging fire season is like going to war.

Veteran warriors—Navy P2B Neptunes, PB4Y Orions, KC-97s, DC-6s, DC-7s, C-130s—lumber in over the hills and drop into the airport west of town. From underground tanks, they suck up another load of a red slurry, a fire retardant, to dump at the center of the flames. Helicopters swarm as gnats beneath the bombers. The choppers trail ninety-gallon buckets at the ends of long cables. They fill the buckets by dunking them in nearby rivers and lakes, then they dump the water on hot spots, stubbornly burning snags, or logs that have rolled down a canyon wall to expand enemy lines. Twin-engine de-Havilland Otters drone from the runways, each carrying nine or so smoke-jumpers, the elite first-strike troops who parachute to initial wisps of smoke.

Later the buses and trucks ferry the troops who wage the ground war. People in fire-proof yellow shirts and green pants swing tools called Pulaskis, a crossbreed of an ax and a hoe. This is the weapon of choice in the weeks of hand-to-hand combat a major fire will require. Command tents and cooks tents spring up. Phone lines and satellites link computers and voices back to Missoula. Weather data pours in from remote stations. Terrain and cover are fingered off on topographical maps. Expert guesses register and flacks arrive to brief the press. So many thousand acres already burned, so many thousand more by Tuesday if all goes well and Missoula can send more dozers, fire lines on the whole perimeter by Saturday, control by Monday unless the rain doesn't come.

As the fires spread that summer, the scenes multiplied. First a few hundred troops were in the field, then a few thousand. By the end of 1988's hot season, more than fourteen hundred separate fires had spotted the northern Rockies. The fighting of these eventually conscripted more than fifteen thousand troops to lean on the supply web centered in Missoula. Back in town, everyone got a piece of the action. Dry-cleaning shops had mountains of yellow sleeping bags by the back doors. Rented trucks queued at grocery store docks hauling off caselots of everything from canned tomatoes and Hershey bars to Tampax. Catering trucks were in a constant frenzy. The Forest Service leased every bulldozer it could find. A perpetually broke and thirsty rugby team rented its ancient team bus to the firefighters. By summer's end, the team had a ten thousand dollar certificate of deposit banked against the cost of future kegs. Wits accused the head of the chamber of commerce of carrying a can of gas and some matches in the trunk of his car, and he'd laugh at the joke as readily as anyone. Such are the effects of prosperity. Everybody worked, from trained Forest Service crews flown in that summer from all states and Puerto Rico to every college kid fit enough to step on and off a chair fifty times. Magnum paychecks were issued to anyone willing to spend the summer sucking smoke and sleeping on the ground. At one point, fire officials started emptying the prisons. I saw crews of tatooed convicts marching toward the lines, violent men wielding Pulaskis. If fires could be won through intimidation, that scene surely would have turned the trick.

It was my job to report this news. It was of considerable interest in towns where skies glowed red each night from not-too-distant fires. I was having almost as much fun as the prisoners, freed from my desk and functioning as what I imagined was a reasonable approximation of a war correspondent. It was a diversion especially welcome because it came at the same time the Missoulian's management was bickering about the publication of my timber series, a good time to get the hell out of the office. I came to work in jeans and hiking boots every day, looking always for a new fire worth seeing, some new angle fresh enough to distinguish the latest inferno from the several hundred that had come before. Often I would find such an excuse, grab a photographer, and roll off cross-country in high reportorial dudgeon.

Early in the season I had stolen from the Forest Service a yellow regulation fire shirt, a hard hat, and a fire shelter. This last piece of equipment shows up as a shoe-box-sized yellow nylon case always clipped to a firefighter's belt. It contains a folded foil pup tent that is the firefighters' shelter of last resort. When the flames threaten to overrun him or her, the firefighter unfolds the shelter, climbs inside, and hugs the ground. Then the hope of a singularly focused brain is that the inferno will run on quickly before its heat bakes the firefighter or its oxygen-sucking flames snuff the lungs and heart inside the tent. I did not believe I would ever need this equipment, but wearing it was a badge that would get me past roadblocks and newly deputized roadblockers. I wanted to be behind the lines. This strategy led to a close call only once, when the wind shifted and a line of fire came roaring straight at me. I escaped easily but it was tense for a few minutes. Still, the slight risk was worth it in that it got me close to the story. Photographer Michael Gallacher, working with me on that fire, looked up just in time to see the aluminum belly of a slurry bomber barely clear the trees, then loose a load of retardant like a great crimson rooster tail. Retardant stuck to his lens afterward.

All that summer I watched the flames work, sucked the same smoke as everyone else, listened to people tell stories about the night before when the fire would hit the crest of the pine and move across a draw faster than they could flee in pickup trucks. Toward the end of summer, a pattern emerged in these conversations, especially conversations with the veteran firefighters: they had never seen anything like the woods of the northern Rockies that summer. The can-do pronouncements and commands to subdue nature were left to the politicians. The opinion of the firefighters was nearly unanimous and humbled: these fires were going to do what they wanted to do; the control of any one of them only came through a combination of lucky breaks, through countervailing forces of nature like a quick dump of rain or a shift in the winds. These fires behaved as they pleased.

Our people have been quelling forest fires in the West for nearly a century. Throughout the region, this has allowed limbs, blow-downs, dead trees, and litter to accumulate on forest floors. To a firefighter this is simply fuel, and so our efforts have not protected forests. Instead, we have made a trade. Nearly a century's worth of small fires have been snuffed in exchange for the inevitability of really large fires now. We thought we were protecting forests, but we have turned them into great bombs, the fuse of which is connected to cycles of drought.

This bomb imagery, however, describes this situation only in terms of its threat to us. We mostly worry because these fires now will spread beyond control, leveling houses and towns, wiping out vast tracts of timber on which the region's economy depends. That is but a small side of the problem. Our focusing on it is merely illumination of our anthropocentric bent. There is a deeper problem: removing fire from natural systems that evolved in symbiosis with fire greatly undermines the health of those systems. That is, in our great rush to preserve the trees, we have killed the forests.

MY UNDERSTANDING OF THIS ISSUE DID NOT EVOLVE, DESPITE its notoriety that summer, in the fires of Yellowstone National Park. The fires there I regarded as more a circus than a lesson, especially a circus political. Those fires, as if by some impish design, demonstrated the ability of America's sound-bite culture to warp subtle issues completely beyond recognition. An incident involving the national press corps comes to mind.

The event was Michael Dukakis's campaign swing through the then-still-burning park. An entourage of reporters had swooped in with Dukakis on two airliner-sized jets. Many reporters, like the candidate, were shod in spiffy new boots that were some upscale department store's ideas of what one ought to wear when hiking. Never mind that the entire entourage's exposure to the wilds on that trip was framed by a bus window, that the rugged trails they trod were blacktop parking lots stalked by Winnebagos. This was theater. There were actual trees, trails, and flames to be photographed as backdrop for sound bites on the evening news. Against this backdrop, candidates and reporters must wear appropriate symbols.

At the center of this symbolism circus was Sam Donaldson. Through his day of shadowing the candidate, Donaldson served ABC News not in boots, but in a blue blazer, at least until it came time to record and bounce his reportage from a waiting satellite. Then an aide to Donaldson appeared and handed him an insulated nylon vest for that properly outdoorsy look on camera. This is journalism in America, and thusly flowed the symbols from Yellowstone the whole summer long. Tape and photo editors thrilled to the lurid images of flames licking away at a national treasure. At one point, a correspondent for National Public Radio tagged a story by claiming to be reporting from "what is left of Yellowstone National Park." Full-color coffee-table books issued. School kids took up collections of their milk money to save the animals. Someone from the Deep South mailed along a box of fair-weather species of pine with instructions that park rangers use these trees to reforest the frigid park. Politicians howled against the natural fire policy, which lets fire resume its natural role in parks and wilderness areas. Never mind that Yellowstone officials were by then fighting the flames with every ounce of power they had.

In the end, none of this, not the politicians' demands, the sound bites, or even the firefighting mattered. The ecosystem we call Yellowstone Park simply went about its business. It demonstrated without doubt and against all human demands to the contrary that its time to burn had come. This is how Northern Rockies ecosystems heal themselves. It burned about a third of its area by the time nothing more technological than a blanket of winter snow quelled the season of fire.

Still, in this rage of inflamed political commentary, few of us were left with any information about the significance of the serotinous pine cone. Too bad, because in this tight-fisted little pineapple of a seed pod nature explains itself. Nature clearly states its intent for the lodgepole pine. This cone is our first grip on fire.

The lodgepole's cone is different from all the rest of the conifer cones, even to the untrained eye. It is closed tight and sealed, no gaps to free the seeds inside. What's the point? It seems analogous to a rancher trying to ensure the propagation of his cattle by putting condoms on all his bulls. Yet in a lodgepole's life there comes a triggering time. One need only spray flame under a serotinous—the word means a delayed reaction—cone to watch it blossom like a rose. Then the seeds come slipping out to colonize the world as it stands after the passing of the flames.

Clearly, evolution has hatched a niche for the lodgepole, and that niche is wholly carved by fire. Lodgepole are the colonizers of open ground. They tend to cover huge tracts of disturbed land, usually places stripped by fire or where all the trees have otherwise died. Serotinous cones also will open after prolonged exposure to harsh sun as one would find in most any clearing. Lodgepole, though, are by design an intermediate step, what biologists know as a seral species. Once they establish the covering shade, subsequent species such as Douglas fir move in. Then shade from the competitive trees kills the lodgepole. Fire clears the dead lodgepole, thins the subsequent species, and the forest carries on. In Yellowstone, though, fire has not enjoyed such liberties, and so the dominant vegetation had become seemingly endless stretches of unthinned lodgepole pine. Hereabouts it is known as "dog-hair" lodgepole for its tendency to grow straight and thick like the hair on a dog's back. If fire does not pick its way gently through these stands as they die, more of the dog hair sloughs and matts on the forest floor as an accumulation of fuel. Later in a stand's life, fire comes anyway. Later, though, there is nothing gentle about it. That's what happened in Yellowstone.

At first it seems inherently unfair that we reporters and politicians criticized the Park Service for the results of all this. What happened in Yellowstone that summer was after all, only the result of a failed policy of a century's making. Earlier we, the public, demanded that the park preserve trees at all costs. Such was our bias for life and green. We made Smokey the Bear a cultural icon, but we also have made great firebombs. The Forest Service and the Park Service understood this first and began to do something about it. Throughout the eighties, they attempted to defuse the bomb with deliberately set small fires and natural fires, but this effort came too late. The drought came, the bomb went off, and we, the public, asked for heads on platters. In a way this seems unfair, but not completely. Both the Park Service and the Forest Service brought this on themselves by underestimating the force and scale on which nature operates. The fires of 1988 were a shock, even to the people who thought they knew fires best. Only quietly in the aftermath is there developing a new appreciation for the sheer scale and subtle intricacy of the great sweep of death.

My touchstone for this is not Yellowstone, but a fire called Canyon Creek, the fire I know best. I did not really get to know it as it burned simultaneously with Yellowstone, although that was a truly spectacular event. I have come to know it since—during two years of walking in the at once frightening and comforting bit of earth this fire once covered.

BEFORE 1988, THE CONVENTIONAL WISDOM CONCERNING forest fire was invariably couched in the term "mosaic." That is, in its efforts to sell to the public the natural fire policy, the "let-burn" policy, the Forest Service spoke of gentle, creeping fires. Lightning would strike a dead tree and a fire would fizzle around for months, infusing its grace through the forest. No roar and rush of a wall of flames, just a creeping underburn to clean up the dead stuff, to thin stands of trees and, here and there, to kill small—say fifty-acre—patches of trees. These patches, as they laid out on a ridge, came to be called the mosaic. Everyone agreed they were the very thing for a forest. They produced variety, a polyculture, edges between forests and grass, forage alternating with cover for wildlife, fresh growth of brush, natural succession. Fires allow nature to play across its full range of options.

Certainly, all of this can happen, but contrary to what the Forest Service PR videos of the time told us, contrary, I think, even to what everyone thought, natural fires do not always work this way. The great screaming inferno called Canyon Creek taught us that.

On June 25, 1988, lightning hit a tree standing in the Scapegoat Wilderness. The Scapegoat borders the venerable Bob Marshall Wilderness, which wraps across the Continental Divide just south of Glacier National Park. When fire began there, Yellowstone was not yet an issue. It was still early in the season, still relatively wet. The Scapegoat is wild, officially designated wilderness. It and the Bob Marshall and Great Bear wilderness areas make up 1.5 million contiguous acres of untouched, unroaded lands managed under the strict provisions of the 1964 Wilderness Act. Clearly, this bit of a lightning-sparked fizzle was a candidate for the natural-fire policy as it applies to wilderness areas. Shortly after the fire began, officials of the Lolo National Forest made the call: let it burn.

It did, but at first not much. In fact, it behaved very much as those PR videos said it ought to. The lightning-singed tree toppled and rolled down a rock face that fenced it in a small area between the rock and a nearby creek. There the fire smoldered for about one month, ranging over less than an acre. Then June dried to July, some fortuitous winds intervened, and the fire jumped the creek to begin behaving more like a force of nature. It slipped up slopes, finding fuel on drier reaches of the forest, but still it worked in mosaics. It fit the design for natural fire and so, officially sanctioned by the Forest Service and openly cheered by wildlife proponents, it burned on. It did its work across 60,000 acres between late July and early September.

Then on the night of September 6 Canyon Creek exploded. To see what happened then, we must move to a new vantage, away from the wilderness, away even from Montana, to the view of a satellite's eye framing all of three Western states, parts of six others and parts of three of Canada's provinces. In the center left of a photo of this view, Yellowstone's fires are obvious. Across Wyoming, a fan of smoke filters and fingers, two hundred miles wide and spreading by the time it hits South Dakota. Yellowstone was burning hard, but normally. Canyon Creek was not at all normal, a fact clearly evident in the pattern of its smoke. In the photo, that fire's plume braids straight and tight, twists like a contrail, one thin, dense line compressed by the power of some phenomenal winds. The smoke plume invades South Dakota's airspace no wider than when it left the Scapegoat four hundred miles to the west. This pattern is direct evidence of an anomaly unique in the recorded history of forest fires. What appears to the satellite's eye as stark detail, however, must have appeared to eyes proximate to the fire as stark terror.

The event the photo documents is known as a reverse wind profile or shear. It is a rearranging of the usual layering of the jet stream. Normally, the jet stream's currents move faster at higher altitudes, but during a reverse profile, a river of very fast moving air sinks low. This is a fairly frequent occurrence, but when a wind profile inverts coincident with a forest fire, all hell breaks loose. The low, fast winds at precisely the altitude of the fire act as a horizontal chimney, a draft sucking oxygen through the flames at unheard of rates. The effect, aside from a tight smoke plume, is cataclysmic.

Between late June and September 6, Canyon Creek's normal progress had chewed for itself a perimeter that enclosed sixty thousand acres. Finally that progress had extended outside of the wilderness area on the south edge of the Scapegoat, and so Lolo officials ruled the fire "out of prescription." Well before September 6, Forest Service firefighters had begun to attack. Their efforts, however, seemed puny indeed when the jet stream flipped. In a single night, the fire reared up and roared more than thirty miles straight east, vaulted the Continental Divide, spilled down the sharp eastern slopes of the Rockies, escaped the wilderness boundary and fled on into the plains near the ranching town of Augusta. There this fire died, out of the trees and unable to gain a foothold in the drought-parched, cow-stripped grasses of the plains. In that single night, with flames moving faster than anything living near them can, this fire spread from 60,000 to 250,000 acres, quadrupled.

No humans died, but almost everything else in the fire's path did, including some rancher's cattle. Buildings and fences burned and so a small political fire ignited, but nothing on the order of the controversies surrounding Yellowstone. Forest Service officials were not at all in the ass-covering mood that seemed to prevail elsewhere. In fact, in the immediate aftermath, they seemed more shocked than defensive, stunned, white-faced, humbled and silenced by the raw power of that fire. Orville Daniels, supervisor of the Lolo National Forest, didn't try to duck responsibility for what had happened. Instead, the official line began to emerge, and this line was reasonably faithful to the best understanding of fire at the time: letting fire burn was indeed a gamble, but most of the time events will go according to plan. Between 1981 and 1988, the Forest Service had allowed four hundred fires to burn under the natural fire policy. Only nine had grown beyond predictions. Only Canyon Creek had exploded to an inferno. Nature is capricious, freaks occur, and so we live and learn. Canyon Creek was simply a freak, probably made worse by a history of fire suppression. In the years after the fire, however, we learned much, including information that directly undermined the official line.

NOW ANOTHER VIEW OF CANYON CREEK IS IN ORDER. IT IS A quieter time, June of 1989. It is the summer after the fires and after publication of my timber series. I am on assignment. The national press corps has long since left the state. The blanket of snow that pushed peace through this place last fall has finally peeled from the backcountry. I am walking a trail with Gallacher, a gaggle of scientists, their fellow travelers, and a string of horses and mules. I am covering a fire as I had never done, as very few of the throngs of reporters who mobbed Yellowstone last year will ever do. This was not my idea; it was the Forest Service's suggestion. We head up the North Fork of the Blackfoot River, planning to walk seven miles into the Scapegoat Wilderness and into the heart of Canyon Creek. I have only imagined the scenes we will encounter, images derived from my understanding of satellite photos and accounts of total acreage, intensity, and so forth. I have preconceptions that match not at all what I will see.

The fire had burned all the way to the trailhead, so we walk in black terrain the whole way. At first, the scale of this doesn't sink in, here on the fringes of the fire, here where it was fought. It looks like most fires I have seen, small trees all black and dead. The older generations fared a little better, blackened stumps, brown, dead tops, a few giant trees that will survive. Here and there, beyond a lucky rock face or some other natural fire break, there's even a patch of green. But as our trail winds from the base of the draws to ridgetops and into the fire's center, our view opens to mountain vistas. The enormity begins to intrude. We can see for miles here, straight up the canyon to clear panoramas of the faces on either side. Only death is visible. Toward the bottom of the draw, what was once a carpet of lodgepole is now a seemingly endless phalanx of black spars jammed in a field of ash. At the ridge tops, killed subalpine fir stand stripped and ivoried, trees chewed of charred bark by last winter's ice-toothed winds, a parade of skeletons lined past the end of my sight, over the next ridge and the next, on for thirty miles or so.

As we proceed, it becomes clear that there is no mosaic to speak of, not anywhere inside this fire's fringes. No edges, no mix, no life to balance all this death. There is only the omnipresent sweep of the hand of an apparently angry god. I am stunned by this specter. I am human, while the hand of nature is not. Nowhere but here can this be more clearly demonstrated.

My guru on this trip is Jack Losensky, a Forest Service ecologist. The evidence of death notwithstanding, he grins often, a reaction not nearly as perverse as it first sounds. Early on I find him stooped in what looks to be a desert of ash, poking around. He spots a bit of green shrub. Its siblings are everywhere, seemingly the only species of plant to brave the ash this spring. Losensky identifies it as spirea, a variety of the same shrub my grandmother grew in her yard. He says its presence here is one of the more amazing and unexpected discoveries in the wake of the fire. It seems a common enough plant. Why amazing? Losensky says it is because spirea was exactly the same plant that grew here last year, every year before the fire. This simple fact sets ecology's seminal maxim on its ear: an overwhelming catastrophe has, at least to some species, made no difference.

Traditional ecology holds that catastrophe is the engine of succession. A fire produces a blank slate, which is colonized by early or seral species. These in turn slowly change conditions, paving the way for subsequent or climax species. The system achieves apparent equilibrium, the state of order we humans seem to value. How can the spirea be both climax and seral? It was here last year when this system was at equilibrium and here this year when the system is not. Losensky says it is meaningless to talk about climax and equilibrium in the northern Rockies, because this place has evolved under the hand of fire. The balance here is a deal cut with fire, and evolution has favored an array of plants and animals evolved and adapted not only to survive but to thrive on that deal.

A bit later Losensky stops to poke in the ash again. He's found a saprophyte, a small plant that survives on decaying organic matter. Saprophytes are common, but he can't identify this one. This amazes me, because Losensky can identify everything that grows. He seems to be speaking Latin half the time. His drawing a blank now, though, is probably not his fault; likely this particular plant has not been seen above the surface for maybe a century. Losensky tells me there's a lot of this sort of resurrection going on. Scores of plants are so tightly adapted to fire ecology that they simply do not grow unless there is one. Our policy of fighting fires has forced them into exile, underground. I wonder about the services these plants in turn are meant to provide to the system that supports them. What damage has been done to the system by their absence? Maybe in a decade or so we will begin to understand.

The year after the fire, Losensky and his staff began compiling a list of endangered plants in the region. As biologists began looking, they located those scarce plants in odd places: around logging sites, in road cuts. These plants evolved to thrive in the wake of fire's upheaval. With fire gone, they began to cast about for an approximation of catastrophe and hit upon the tracks of bulldozers. Forest Service scientists don't like to talk about this particular bit of information, knowing full well how it will be seized upon in the political upheaval of the state. Already loggers blast the natural fire policy, claiming their clearcuts approximate the effects of fire, so why "waste" the trees? Why not let them log the trees instead of burn them, if the ecosystem would be as well served? Why not let loggers provide dozer tracks for the endangered plants?

It's a fair question outside Canyon Creek, but immediately ludicrous here in the aftermath of the fire. To begin with, fires do not need roads, dozers and trucks. Some species may carve out a make-do milieu in roads, but the roads are still the greatest damage from logging to the overall system. Beyond roads, though, there is an even more obvious dismissal of the loggers' contention: forest fires do not burn trees. Forest fires only kill trees. The distinction is hugely important. The tree's charred body remains, usually standing, to be attacked by bugs, then birds, then decay, to lie down and rot and fuel life. We miss this single and monumentally important fact because we are conditioned to avoid looking at death. To a forest, the remains of a dead tree are as important as its life. Our greatest sin against the forest is not killing trees. Nature kills trees with abandon. Our sin is to cart off the bodies.

This reckoning with the effects of fire and logging begins to nudge me now. Beyond its particulars, fire begs a universal and long-nagging question of environmentalism: how do we define "right" in terms of our responsibility towards the earth? Nature kills. We kill. Nature adapts. Given this, is it possible to wrong nature? Fire suggests a beginning step in tackling this question, or rather, a necessary preliminary question: what do we mean by "right?" It is a question of values, and nature is valueless. Nature does not welcome or abhor catastrophe. Nature adapts to catastrophe. Nature is not right or wrong. Nature is, in its essence, adaptation. The creator is evolution. Does this obviate, then, the question of right? Not really. To clarify the question, though, we need to understand that it springs from a human value.

We are not separate from nature. We are adapted to the web of life as it exists. As a species with the ability to profoundly alter the web, it seems we ought to understand the effects of our alterations. Nature's adaptations are so sweeping and patient as to threaten our own survival. Nature is complicated, though. We never can understand all of the consequences of all our actions. In this uncertainty, then, the least threatening situation is a maintenance of the status quo, to change nature as little as we are able. We cannot know whether one of the dominoes we topple will ripple back through the chain to fall on us.

What is right, in human terms, is what has always been, at least in the sense we understand "always." What is right is the finely tuned balance nature has achieved since the last catastrophe like the Ice Age. We fought fires because we believed that they were wrong because they killed. What we did not understand is a finer notion of killing, that the death of the individual is necessary to ensure the survival and vitality of the species. We did not understand that in the natural sense (ultimately in the human sense) fires are right.

Nature, through the random work of evolution, creates an array of life to deal with conditions as they are. If that condition is fire, and if it exists for twelve thousand years, life will eventually come to thrive and depend on fire. We cannot tip that balance without consequences. Humans do have the power to foist upon nature upheavals such as logging or fire suppression. Nature has no bias against these upheavals and will adapt by trial and error. Through evolution. Through wiping out a huge array of species that no longer fit the new rules and bringing on a new set that will. Ten millenia hence we may approach equilibrium again. Then humans may be one of the species here to observe this new order, provided we are not among the obsolete wiped out in nature's rush to adapt. We need to understand that the notion of right and wrong hinges on our power to create the next global catastrophe, rivaling the upheaval of the Ice Age. Nature will not prevent us from doing so. Nature will only adapt after, in human terms, it is too late.

ON DOWN THE TRAIL, A SEVEN-MILE WALK FROM THE NEAREST road of any sort, we come to a Forest Service cabin where we'll stay tonight. Dick Hutto, the University of Montana ornithologist, is already here. He spotted a grizzly earlier, but he's here mostly to do bird work. He has a grant from the National Geographic Society to study the effects of fire on small birds. To this end, Canyon Creek is the happy hunting ground. He's flapping and hopping around, clearly elated by this forest of char.

"Not dead at all. Come here. Listen. You can hear them." And we do, a group of a half dozen people standing, ears cocked to black snags. Hutto shows us how to find the good listening trees: Look for bits of wood, like fine sawdust, at the base. We find such a tree and listen again. When we talk, the tree becomes silent. They hear us, so we hush. Presently comes the "ritchy, ritchy, ritchy," the crunch of beetle larvae boring through wood. Hutto becomes more animated still. "See, you can hear them."

They are a species of pine beetle evolved to sense fire. The process is thought to involve detection by infrared light, or some other arcane mystery of bug communication. These beetles detect fire from miles away and move into a burned forest even while it still smolders. If there are no fires for years in a region, then these bugs have a tough go of it, surviving by attacking healthy trees. Still, they prefer the black.

Why is an ornithologist so intent on bugs?

"This is just how they find them." (The "they" this time being woodpeckers.) "They hear them," Hutto says.

Then Hutto's eyes kind of bug out. He folds at his waist to arc his head in a swoop. "Like this." And then he bangs his beak into the tree right where we heard the beetle larvae at work. (I work with bird guys with some trepidation. They may go weird at any time.)

Hutto's bird is a black-backed woodpecker, evolved every bit as intricately as the beetle. He shows me how to identify the birds. They can be found outside of the black. Rarely, though, in concentrations such as we see here in Canyon Creek. Hutto has been studying this business his whole career, yet he tells us the complexity of this arrangement never really hit home until he saw the woodpeckers at work here, flitting incessantly among the charred lodgepole. It makes sense. It's obvious really, pure black backs that are unique in the woodpecker trade, black so they are invisible working a fire-charred tree, invisible from predators. A fire bird.

The woodpeckers will remain here only a few years, then move to more freshly burned trees. They will leave holes that are ideal for the nests of bluebirds. Bluebirds are in trouble in Montana. No nesting sites. People have taken to building boxes and hanging them on fence posts to help the birds. It seems the bluebird is a fire bird, too.

Back at the camp the pack string of mules with our gear has arrived, so we open the cabin for business. This takes some effort, because the front door is armored against a sort of storm unique to this area. Every couple of square inches holds a long, steel spike, protruding point first straight out from the door. This is learned behavior for the Forest Service, doors designed to keep grizzly bears from looting the cabin. Once laid down, this device looks like something a Hindu swami would use to test his faith. I tell the Forest Service guys they didn't have to go to all that trouble just to make a bed for the reporter. They like this idea.

We stash our gear, then I ignore all distractions to concentrate on business at hand. I noticed on the way in that the ash around the cabin had urged on a bumper crop of morel mushrooms. This species of fungus exists throughout the country. My woodsman grandfather in Michigan was a morel hunter of some conviction and taught me the art. Oddly, though, the particular variety of the northern Rockies is wildly stimulated by fire. One hardly ever sees them in the unburned forests of this region, but the first few years after a fire, they proliferate lasciviously. Huge morels growing shoulder to shoulder. At times, one can't plant a boot without stomping one. Gallacher, also a morel stalker, and I hit the woods like kids at an Easter egg roll. We vacuum up fist-sized morels by the hat-full. The forest would not miss these fruiting bodies. I remind myself to carefully attend to the disposition of the spores in the morning.

The Forest Service people know nothing of morels and so are a bit skeptical when Gallacher and I return lugging booty. Still the mushrooms all disappear as I spend the long, near-solstice evening running fungus through butter and a cast-iron skillet. Fed by the forest, Gallacher and I unroll our sleeping bags on the cabin's back porch to wrestle with a full load of rich food and bear dreams, a staple of backcountry nights. These dreams are not at all helped by the cabin's resident mouse, a comedian. The mouse is greatly entertained by crossing and recrossing my sleeping bag. I keep winging chunks of firewood at it but succeed only in hitting Gallacher. Finally, I give up and fall asleep. Then the mouse liberates for its nest huge bites of the Ragg wool sweater I am using for a pillow.

AT FIRST FLUSH OF DAWN, WE LIGHT OUT FOR THE RIDGES, JUST Gallacher and I. Presently he splits off in search of new patterns of illumination bending through the carcasses of trees, a photographer chasing sun like a springer worries pheasants through the brush. He disappears. I stand alone at ridgetop about a mile above the cabin, sole witness to a city-sized vista of death. This solitude proves to be prime habitat for human ruminations, clinging as they do to life and green. I wonder about the forest, when it will return. By now I understand the importance of the fire to the eventual vitality of the forest. Still, I do not consider this bleak scene before me to be a forest. The day before, I had seen a patch of land that had burned the previous decade. It had reseeded itself and ten years later, shoulder-high saplings jostled each other for a piece of the sun. Ten years, and still no forest. What of this place? When will it be forest? In my lifetime? There it was again, that phrase: "In my lifetime." I had used it repeatedly in Canyon Creek, the yardstick by which I measured the "damage." Now it appears as if I am not so much concerned with the damage to the forest as I am with damage to myself. In my lifetime. What does this forest care of my lifetime? Somehow, my sadness for the death of these trees is wrapped in the sadness for my own death.

Wilderness is a grief-stricken place. I have known this as long as I have walked among timeless warps of rocks laid down by forces that consider my life of no consequence. Wilderness invariably convinces me of my own insignificance, even of the insignificance of my time. But here in Canyon Creek the grief tunnels double deep into my center. It is a grief that provokes at first self-pity. How could nature be so cynical as to freight its creation in all this death? So cruel as to equip humans with all of life's will to survive and yet give us a highly evolved and unique tool capable of contemplating our mortality. It seems a vicious joke, giving us a preview of our deaths as they are mirrored in the death of those lives around us. But then nature really isn't vicious. Nature isn't kind either. Nature just is.

Light interrupts my thoughts. Just now rises the sun that has been blocked by a distant ridge. Light transcends to shoot through the tree skeletons that line the horizon. This fire-drawn place accepts this light and now seems transformed in my eyes. The sun shifts my head a quarter turn. I abandon my demand of this forest that it masquerade in green. Now dead and white in the morning sun, this forest blazes with a beauty of its own. It is dead, and it is life.

I think of a story Losensky told me while we walked along a trail. Near a creek bottom, a patch of trees stood still green, survivors of the flames. Losensky said this was not at all unusual, that because some species can live in wet places, individuals can use the creek bottoms to hide from fire. Stands of other species are purged by fire as routinely as every fifteen years, but some stands of, say, cedar or spruce, do not burn for maybe four hundred years, apparently exempted from nature's need to create with death. Only apparently.

Nature has backup systems. When the lodgepole pine reaches a certain age, it is alive, but moribund, can continue in this state for centuries, serving only to suppress the coming generation. And it would, were it not that when the species crosses this line, it begins to emit an audible signal. Audible. Then there is a species of beetle so finely evolved as to hear this signal. It is attracted to the moribund trees. It attacks and kills them. This is how exquisitely attentive nature is to death, its creativity. The forest is so finely tuned as to produce a tree that sings its own death song.

LATER THAT DAY, I CORNER LOSENSKY SO THAT I MIGHT PLAY reporter again. There have been too many hints here, too much evidence to contradict the official party line about this fire's being a freak. This fire seems to fit this place, not an interloper at all, but a completely native species. Losensky lays the whole business out bluntly. True enough, there are places where natural fires occur gently, mosaics and under-burn, but this section of wilderness is not such a place. The record is clear on this. In fact, when stacked against others, the

Canyon Creek fire was something of a middling effort. Between 1889 and 1926, fire burned more than half of the 1.5 million acre Bob Marshall complex, of which the Scapegoat is a part. The famous fires of 1910 set 360,000 acres of this vicinity ablaze, half again as many as Canyon Creek did. Losensky eventually dug up an 1899 survey made in preparation for adding the land to the system of national forests. The survey was the work of H. B. Ayers, who spent a couple of years penciling a detailed, color-coded map of the region. But he also spent a great deal of time sifting the remains of a fire that had swept the Scapegoat in 1889. Curiously, that conflagration had covered the Scapegoat in exactly the "freak" pattern of the 1988 fire, roaring out onto the plains near Augusta on September 1. The Canyon Creek fire hit just the same spot ninety-nine years and five days later.

Ayers also found evidence of a similar fire about forty years before that. Losensky says that is the pattern of this place, not gentle fire at all, but great infernos every twenty to forty years. This pattern is a result of a collision of normal cycles and the unique terrain of the Scapegoat. Mostly the fires have corresponded to natural drought cycles, regular recurrences of a series of dry years throughout the West, but some local circumstances amplify the effect. The area is relatively wet and so fails to burn with every drought, storing up larger quantities of dead trees as tinder, a sort of storing of energy, a winding of the spring. The Scapegoat's south edge is a series of canyons that face into the arid southwest winds of late summer, a unique pattern created by the Blackfoot River Valley. The orientation turns the canyons into air vents feeding the chimney effect created by the reverse wind profile.

The coincidence of these factors occurs in a pattern largely written by drought, but this pattern can be read on the land. Based on Ayer's work and that of others, Losensky concluded that historically the Scapegoat has been covered with fire-enforced grasslands and few trees, not dense forests. As soon as a forest gains a toehold, fire sweeps it off most of the land except for the wettest north-facing slopes and creek bottoms. Ayers's maps showed the area about five percent covered with dense forest, likely typical cover from the time the glaciers receded until the turn of the century, when fire suppression began. Just before the 1988 fire, the Scapegoat was eighty percent covered with dense forest.

I think of the phrase "untrammeled by man," which occurs in the 1964 Wilderness Act, the law that governs the care of the Scapegoat. We do not have the political courage to face the consequences of this phrase. We don't mean "untrammeled," we mean preserved according to our biases for green and trees. Nature's biases for this place are something else, fire and death. How much more trammeled can an ecosystem become than to lose its delicate balance between fire and trees, to be pushed from five percent to eighty percent timbered? Since the Ice Age, nature has evolved in the Bob Marshall an entire array of life designed to live among fire, grasslands, and sparse, parklike forests, the system that is in an evolutionary sense "right" for the Bob Marshall. Since the turn of the century, however, we have installed dense forests here because of our biases, and now evolution must patiently adjust to the effects of our heavy hands.

Losensky and I talked of all of this quietly because we both know the political implications of these facts. It is an issue that goes well beyond the usual cleavages between developers and environmentalists. Many environmentalists are simply preservationists. They are not ready to accept this much nature, raw and uncontrolled. The consequences of this fire are controversial, but then does it really matter what people are ready to accept? As Losensky and I walk and begin to let our minds mix the science and politics, there emerges a realization that the politics do not matter. The cycle of fire has done its work. It will be another forty years, probably, before this issue emerges again. By then what

Losensky said and what I reported will have been forgotten. Our political discourse forgets what happened last week, never mind what happened forty years ago.

But even if the public does remember that the Bob, left to its own devices, burns like a bomb, so what? If the political powers resume fire suppression, it will only delay the inevitable. A century of fighting fires, in the end, made no difference to Canyon Creek. True, there was no fire of this magnitude in the late forties, a synthetic gap in the forty-year cycle, but this, in nature's time frame, is but a momentary skipping of a beat, the skipping that allowed an unnatural proliferation of trees, of fuel. In the end, fire comes. Perhaps we can alter the schedule a bit, but the ultimate outcome is beyond our control. Nature is patient. Sooner or later, a drought cycle and a few bolts of lightning will align randomly with another inverted jet stream, and the canyons of the Scapegoat will feed an explosion. This place will burn, no matter what we do. Nature will right itself and begin whatever it takes to adjust to the damage. Fire is as inevitable here as drought and wind, as inevitable as our own deaths.

In the end, there is no possibility of control here in Canyon Creek, which makes this place, in a sense, inhuman. At times it seems as if all of human inquiry is fenced by our need to control, our greatest bias. In recognizing and removing this bias, there is liberation. The need to control is rooted in fear. The acceptance of the power of fire, a surrendering of control, is freedom from that fear, the fear that is the mirror image of the fear of our own deaths.

Now, years after my visit to Canyon Creek, this lesson of fire seems to want to cross from the forest into my life and into all the new places my questions choose to explore. The urge to rule fire is but a particularly obvious example of our hubris, but hubris is everywhere. Fire comes round every forty years or so, often enough to teach. Yet the same ineluctable hand of nature rules all. Everywhere, the notion of our control of nature is an illusion. We cannot control, not in the long term, not in the end. Our attempts to do so, however, do great damage, ultimately, to ourselves. The illusion of control survives only within that tight circle circumscribed by our own narrow questions. The illusion explodes in flames when we venture beyond that circle to broader, more satisfying and more unsettling questions.

CHAPTER SEVEN

WHERE THE MOUNTAINS confront Montana's plains, geology has drawn a clear line against human intentions. Here, continents collided to produce the folds and lifts that are the Rocky Mountains. The massif's stunningly abrupt eastern edge is fortified by formations called reefs, sheer walls of rock that check the flow of civilization. To the west of this border lies a broad band of wilderness, breached here and there by highways crossing the few passes over the great divide. To the east lie cities, ranches, railroads, and farms. It is the edge of our world. Despite the exceptional incursions, mostly this front has held. Yet it seems as if we humans are not unique in our butting against this line. We are but the latest species to follow an ancient pattern dictated by the confrontation of continents. It seems as if this line traces, as some people believe, the fulcrum between the weight and counterweight of creation.

Montana Highway 200 is a two-lane asphalt strip that winds from Missoula upstream along the Big Blackfoot River. It trends more or less east through a tight canyon, a slit in the mountains. Then it opens to a cow-specked valley that holds the burg Ovando. Occasionally, grizzly bears wander into town from the Bob Marshall to the north, but this is rare. Mostly the town entertains itself on its quiet eccentricities and in Trixie's Antler Saloon. From Trixie's, the highway winds higher, still along the Blackfoot and just south of the Scapegoat Wilderness, to Lincoln, a loggers' town. Senior ponderosa pine mind the highway's shoulder. Deer stray into restaurant parking lots. Rogers Pass, the highway's notch through the great divide, is but fifteen or so miles east of Lincoln. On January 20 of 1954, it was seventy below zero at the pass, the coldest temperature ever recorded in the contiguous United States.

From the pass, the highway drops quickly to the plains. The ponderosa are gone now, replaced by a few Douglas fir, growing sparser still as the highway descends. That line between mountains and plains approaches; east of it the forest ends. This is the route I took to leave the forest one August afternoon, on assignment for a change, not writing about trees. Just after Highway 200 rolls into the plains, my Jeep cut north on 287, past Hutterite colonies and sheep ranches. This is an arid fringe of the Great Plains, barely arable. Mostly it is cattle land, not at all like the squared and rowed wheat fields further east in Montana, in the Dakotas and on into Minnesota. Here driveways are often ten miles apart. What few towns there are always seem a long ways away. I drove north toward Augusta, near the spot where Canyon Creek's fire found these plains, then to Choteau, the next cow town up the line. The assignment, however, had nothing to do with cows. I was seeking evidence of dinosaurs, plant eaters that had grazed here in other millenia. More accurately, I was hunting young urban professional people who appeared to be victims of a scam.

This scheme was the work of Dave Swingle, educational director for the Museum of the Rockies in Bozeman, Montana. That institution has taken a considerable boost in its reputation through the work of Jack Horner and his late

partner, Robert Makela. They were a couple of Vietnam vets
attracted to the wild and rough life of paleontology in the
seventies. Long on iconoclasm and, at least at first, short on
training, they nonetheless revolutionized the field. They are
credited with certain technical innovations, including design
of a paleontologist's backpack capable of lugging a case of
beer. The site of one of their earliest digs was reverently
dubbed "Camp Rainier" in honor of the best brew they
could afford at the time. Other contributions, however, were
more substantial.

Farming a rich delta of bones near Choteau, Horner and
Makela raised the revolutionary notion that dinosaurs were
really more like birds than lizards, were warm-blooded, and
reared their young on nests. Makela later died in a truck
wreck, and Horner has ranged away from Choteau to new
digs in the northern plains. Still, there remain plenty of un-
dug bones at the original site. There, in just a few acres, an
area the size of a small city subdivision, there are the remains
of about 10,000 individual dinosaurs, mostly of the genus
Maiasaura, a 25-foot herbivore. Horner and his colleagues
had unearthed so many of these creatures that further digging was becoming redundant. Still, it was possible that the
remaining bones hid some secrets. That's where Swingle
came in. By the time Horner moved on, the diggings at
Choteau were mostly depleted enough to be entrusted to amateurs. Swingle shanghaied the amateurs. He bet on what he
believes is a quiet undercurrent in the nation. He thinks some
people are so fascinated with science they will pay to per-
form the stoop labor of a dinosaur dig.

He rounded up fifteen tipis of authentic Northern Plains
design and pitched them in a sun-baked, wind-blasted, rock-
strewn stretch of the plains. He chose tipis because he found
they were the only temporary shelter capable of surviving the
violent East Front winds, a tribute to the design's fidelity to
place. Then Swingle placed a few ads in prominent magazines

offering a vacation package. Anyone willing to pay the museum $850 could come to dinosaur camp for two weeks. During that first summer of 1988, Swingle got eight hundred takers.

Photographer Kurt Wilson and I find the latest batch of diggers easily enough, just west of Choteau in the only encampment of tipis around. The vacationers are assembled nearby, looking like a chain gang: teachers, dentists, and computer programmers stand on rock piles making big ones into little ones. It is a sort of over-achievers' dude ranch. I corner Swingle quickly and we pick out a spot in the shade of one of the tipis. He is the sort of interviewee who asks a reporter how much tape he has, then measures out his comments to pack the whole cassette. I decide he has packed his brain in much the same fashion. His monologues play out more like a dissertation or elegant mathematical proof than like conversation. He was a high school principal who taught both metallurgy and creative writing. Now he is a paleontologist of sorts, a museum fundraiser, a historian, the operator of the dinosaur dig's antique Oliver bulldozer, peeler of carrots and the camp's only expert on the care and feeding of its propane refrigerator. Our interview is more a case of his talking and my listening, stunned, as his acrobatic brain tripped through the eons as casually as most of us page through a phonebook. In a full-alarm mental overload, I flee Swingle, hoping I could pull something usable from the tape later.

Now I lock onto my real targets, the suckers who are paying good money to swing a pick and shovel. They are just outside of Swingle's tipi, rapt in menial recreation. I draw a bead on a fat man from Washington state. He is an irresistible target, resplendent in bermuda shorts, loafers, and black socks, a hand lens slung on a string around his neck, K-Mart straw hat, and t-shirt losing the battle to span his middle latitudes. He is exactly the sort I expected here. My target approaches his slight, pale wife who has spent the day jamming an ice pick into the hardpan.

"What did you do, dear?" he asks.

"I found a bone."

"Well, let's see." She shows him.

Elated now, he says: "Well, we'll take a picture of it." He fetches a movie camera to better record the action of something that has not moved in 80 million years.

Elated now, I write this down. This is exactly the sort of silliness I expected from these people. It should fit nicely in the taunting story I have planned. I venture from my target seeking more ammunition from similar specimens spread around the dig. I speak with the diggers. Oddly, I begin to hear them speaking back, but in a tone an octave or so removed from the nerdspeak I expected. An octave lower maybe, tones ringing as awestruck. There are a couple of teen-agers here, one so intense that he won't answer my questions for fear of diverting his concentration from science. He stops digging to try to listen to me, but after a few seconds begins to fidget and shake like a lap dog about to pee on the floor. Finally he ignores me and resumes his digging. He is only the most extreme case. No one seems interested in taking a break from the chain gang to talk to me in anything other than clipped sentences. Their attention is elsewhere. I give up on interviews and instead take an icepick and brush and begin prying in cracks of my own. We are working a garage-sized pit only a few feet deep in the hard shale and clay. The pit is stringed off straight and square, like a picture window into the earth. There must be thousands of bits of petrified dinosaur protruding from the hard pan. These pieces I soon begin to recognize as femurs and vertebrae, bits to which parts in my own body correspond. Now I find I am less interested in talking, more intent on flicking off the dirt covering these bits of ancient life gone to rock. I have a hand on what were once living cells, part of an individual. Its life is separated from mine across a stretch of time I cannot imagine. Its home, however, its place on earth, was just the same as my own. I am touching a creature that died along with ten thousand of his relatives, all within a few catastrophic hours, all the bodies heaped in a single pit.

Later we move from the clay and that square pit to a pile of rocks less than a mile away. The rocks are known to contain eggs. One finds them by gently breaking the rocks or by sifting through the mounds of chips sloughed from the pile. A lucky digger of dinosaurs will encounter a few egg flakes each day, just the flake, exactly the shape, texture, size and thickness that would result if a rolling pin were run over the shell of a hen's egg. I find such a flake, no longer shell, but rock now, but still looking for all the world like an egg, stippled surface and all. This is evidence of the beginning of a single life eighty million years ago.

It was not a rolling pin that made this flake, not even the eons of grinding by these rocks. In fact, the configuration of these shells allowed Horner to deduce the business about nurtured young. In piles such as this he found only bits of eggshell and bones of infant dinosaurs. He found infant bones no where else. From this, we are to understand that the young creatures hatched but stayed on the nest, growing and slowly stomping their eggs to flakes. Because they stayed put for a time, the adults must have fed them. This shell I hold was cracked by the beak and feet of a new bit of life. The rules of the dig are specific on one point. We diggers are not to hoard our finds but must surrender them for analysis. I have another idea. I have only just now established contact with a life from a long time ago, and I find I cannot let go. Surreptitiously, I slip the bit of shell into the watch pocket of my jeans.

THE WORKDAY IS OVER NOW AND THE DIGGERS ASSEMBLE FOR a Swingle-cooked dinner in a specially built tipi that is forty feet in diameter. It is the largest known authentic tipi in captivity. We eat our stew and work on the camp's copious supply of Rainier. The evening's scheduled recreation is a hike through the nearby Pine Butte Swamp. It is a former ranch bought by the Nature Conservancy, a private group dedicated to saving threatened habitat. The organization has preserved the swamp for grizzly bears. Once that near-extinct species roamed the plains over most of the United States, but the advance of civilization pushed them back behind the line. Adaptable, they have converted from plains to mountain dwellers. Mostly their remaining habitat in the lower forty-eight states lies in the high mountain passes of wilderness areas, most notably the Bob Marshall Wilderness and Glacier and Yellowstone National Parks. The most outstanding exception, though, is Pine Butte Swamp, which lies near the Bob's eastern edge.

We walk through the swamp, which is different from any stretch of prairie I have seen before. As its name implies, there is a bit of water here and so it supports exceptional vegetation, a rich mix of grasses, sedges, and forbs seldom seen elsewhere. These are just the stuff for bears, which are, contrary to reputation, eager consumers of plants. A resident botanist is guiding us through the swamp, a fen really. I listen, trying to span at least some of her command of the diversity of this system. Then the sun begins to set, dropping evening like vespers on the place. In the waning light the swamp seems haunted by the grizzly bear. The swamp appears vast. I begin to feel as alone here as I imagine the grizzly must, threatened as it is with extinction.

We return to our tipis. Now it is full dark, with just a bit of a glow still hanging on the fringes of the mountains that wall the western horizon. It is full dark as only places well outside the glow of city lights can become. There is nothing quite as immense as a prairie sky full of stars. The camp's agenda takes full advantage of this show. A volunteer in camp is Bob Yaw, a retired meteorologist and an amateur astronomer with encyclopedic knowledge of the stars. We diggers circle up round Yaw. We find Saturn. He begins ticking

off the constellations, first as we understand them in classic
Western tradition, as the Mediterranean shepherds did, and
then as they appeared to Navajos.

As Yaw speaks, I feel less lost in this day of jolts against
the edges of my comprehension. Here we are now among the
stars, the infinity of which I have confronted my whole life,
almost as if their vastness so far surpasses all else as to provide a reassuring border to the universe. I feel at home in
them, as if I could leap from these rocks into the undiluted
clarity of the prairie sky and wander in the infinity. They define the unimaginable, almost as if all that falls short of the
stars lies within my grasp. Then Yaw arrests my attempt to
flee the incomprehensible. It is nearly midnight and someone
picks up the Scorpio cluster lying low on the southern horizon. He begins identifying the individuals of this cluster just
as I notice the vantage.

The constellation looms just a tick above the square-edged
dinosaur pit, and in this night glow I can see a few tips of
bones silhouetted against the sky. Now Yaw tells us the stars
of Scorpio are young, which strikes me as strange. I have always regarded the stars as timeless, and so the word
"young" does not compute. The specifics are weirder still.
Yaw says these particular stars came into existence a full 35
million years *after* these dinosaurs died. I have touched a life
that is older than the stars.

Swingle has made press quarters in his tipi, but I decline
the offer. Instead, I unroll my sleeping bag on top of a rise
and try to sleep in the noise of these stars and dinosaur
bones. Hours later, a yellow-green band spreads along the
eastern horizon and the tipis begin to glow in anticipation of
the morning sun. I remember I came here to poke some fun
at the tourists. Now though, I am feeling like a tourist too,
impossibly lost in an array of time beyond my ken.

To the west are the mountains, and the ancient line they
draw. Now I realize that the line of collision, of death and

creation, existed here even before the mountains did. Even before the rock slabs slid from the sea that once lay to the west of this plain, this was a temperamental land. The plain likely was lush then, producing enough plants to support a dense aggregation of giant eaters of plants. Just to the west, though, there likely were volcanoes, now covered by the slabs of limestone that formed the northern Rockies. Probably those volcanoes erupted one day, spewing a gas poisonous to life hereabouts and so these dinosaurs, infants on their nests and adults grazing, all died, a local population of a species wiped out in an instant. The occurrence of such a catastrophe, not at all rare in the long view of nature, is really the only way to explain the unique configuration and concentration of the bones.

Just after sunrise, we pack our sleeping bags in the Jeep and leave the dig to bounce back toward Choteau. Along the road this time, we notice an anomalous chain-link fence and array of antennae around a paved lot in the middle of nowhere. Locals take no notice of these places, and usually neither do I. There are hundreds of them dotting the prairie like a shotgun pattern on a map of the area around Great Falls and Malmstrom Air Force Base. They are underground missile silos dug in through the same hard pack that holds the bones of extinct dinosaurs. The extinction we force on the grizzly and contemplate for ourselves is not without precedent. I wonder about the young uniformed people who work in these silos, fingers on buttons. I wonder if they know of the record of extinction all around them.

THE UNIVERSE OF THE BLACKFEET PEOPLE IS AN ARRAY OF EVER tighter circles. They believe the universe as a whole is a circle. Who could argue when confronting it, as the Blackfeet do, from beneath the bowl of a pure prairie sky? They believe their personal universes must replicate creation. That's why tipis are round and oriented always to the east to face the rising sun. When the community gathered in a lodge, the people sat in a circle. The pipe was passed in a circle. The pipe itself was always rotated before it was passed, always clockwise. These circles arc through time. Heritage is not linear, a straight progression from ancestor to progeny, then to now. Rather, there is a real relationship with the past, a give and take, a cycle.

All of these circles became important to me a few months after the dinosaur dig. Again I was on assignment. This time I went further east from the forests, further into the plains, north past the road that led to the dinosaurs and on north along the Front Range to the town of Browning, center of the Blackfeet reservation.

The Blackfeet are unusual in many respects. Unlike most other tribes, the Indian wars did not remove them from their homelands. White oppression instead confined them to a small piece of the land they once roamed, and in their remote home at the north edge of Montana and on into Alberta, they were not finally "pacified" until the end of the nineteenth century. The memories are crisp, therefore bitterness is closer to the surface. Ostensibly, I went to Browning to write about an unfinished battle from those days of defeat, but I found it difficult to address a single topic when dealing with the Blackfeet. Connected still to this imposing landscape of the eastern front, their network of circles seems ever expanding, ultimately enclosing more than expected.

This latest battle had ended in victory. Resting in a Catholic church in the center of Browning was a stack of wooden crates full of bones, the spoils of this war, or more accurately, the reclaimed spoils of an earlier war. The bones were Blackfeet dead, stolen in the interest of science. In the early 1880s, forces of the U.S. Army rounded up the last of the Blackfeet and herded them into the center of their reservation at a site just south of Browning. The buffalo were gone, and the Blackfeet were starving. Their population had been decimated by smallpox. The Blackfeet charge that the army took a lesson from the effects of this disease and began waging biological warfare. The army shipped from the Midwest blankets that had been used by smallpox victims. They gave the blankets to the Blackfeet.

The winters, beginning especially with that of 1882–83, were seasons of disease and starvation. Bodies piled up on the hills around the Old Agency. Ten years later this stash of bones was still growing, and came to be regarded as a "scientific" bonanza. The racist taxonomists then were still collecting specimens of Blackfeet as if they were butterflies to be passed around over glasses of port at the club. The task of collecting these specimens fell to Z. T. Danial, who reports his efforts to his superiors in a letter written in 1892.

> Dear Doctor:
> I have gotten the crania off at last. I shipped them today to Post Surgeon Byne at Fort Assiniboine, in compliance with your request under date of April 6. There are 15 of them. I collected them in a way somewhat unusual: the burial place is in plain sight of Indian houses and very near frequented roads. I had to visit the country at night when not even the dogs were stirring. This was usually between midnight and daylight; after securing one I had to pass an Indian sentry at the stockade gate, which I never attempted with more than one, fearing detection. The graveyard is two miles from the office, so that I have traveled 60 miles on foot to secure them; on one occasion I was followed by an Indian who did not comprehend my movements, and I made a circuitious route away from the place intended and threw him off his suspicions. On stormy nights . . . I think I was never observed going or coming, by either Indians or dogs, but on

pleasant nights I was always seen, but of course no one knew what I had in my coat. I do considerable gunning and am in the habit of wearing my hunting coat about the agency, and you know they have large pockets for carrying guns and I concealed them in one of those pockets. I was always afraid of bringing two, because they would make me look to (sic) big. Sometimes I carried my hammerless and sometimes I didn't, but I always carried my pistol. Nearly every trip I saw wolves who howled at me, they were always near the dead bodies. The greatest fear I had was that some Indian would miss the heads, see my tracks and ambush me, but they didn't. I regret the lower maxillae are not on each skull. I got all I could find. There is in the box a left radius and ulna of a woman with the identical bracelets on that were buried with her. The bones themselves are nothing, but the combination with the ornaments make them a little noticeable.

By way of the Army Medical Museum, Danial's diggings eventually wound up in the collection of the Smithsonian. There they languished in drawers and cartons until a spark in our time shed new light on them. The spark traces to disparate events like the American Indian Movement and subsequent troubles at places like Wounded Knee, South Dakota. In Browning, though, it seems to trace most particularly to the Vietnam War. The longstanding warrior tradition of the Blackfeet, along with the poverty of the reservation, brought many young Blackfeet to Southeast Asia. When the war was over, the veterans faced the same problems as their white colleagues, but their troubles were multiplied by poverty and the alcoholism of the reservation. Some of these veterans eventually sought an antidote for those troubles in their traditional religion. Now there is a more active traditionalist movement in Browning, made up mostly of young and middle-aged men, Vietnam vets, recovering alcoholics, and activists. Again Blackfeet warriors are speaking their native language, learning the rituals of sweat lodges, vision quests, medicine bundles, and the Sun Dance.

My first link to all of this was Curly Bear Wagner, a former AIM activist, the tribe's cultural director, and a large man, taller than six feet. Well before my trip to Browning I had spoken by telephone with Wagner hoping to set up a series of interviews with key players in the battle for the bones. He wouldn't commit to specific appointments, as if schedules and appointments were alien concepts. He only said that I ought to come on up some day and people would take care of me. And so I did, arranging to meet Wagner at Blackfeet Community College in Browning. I was immediately intimidated by his size, his flat, stoic eyes, and his history, which I assumed gave him a justified distrust of whites. Wagner, meanwhile, immediately began telling jokes. He is an easygoing man, instantly friendly. More comfortable now, I began to press him for details of the story about the bones or at least a schedule of interviews. When could I see so-and-so? Who can answer questions? He never really responded, but I got the impression that I ought to keep my mouth shut, that matters would become clearer in time.

Presently, Kenny Weatherwax showed up at the college. He is about Wagner's age. He had joined Curly Bear and a traditionalist delegation that had traveled to Washington D.C. a few weeks before to retrieve the bones. The head of the entourage had been Mike Swims Under, an old man largely regarded as a pivotal spiritual leader. It is through Swims Under that this younger generation has reestablished a direct link with the ceremonies of the past. In Washington, the delegation built a sweat lodge on the banks of the Potomac to conduct a vital purification ritual before they handled the bones. Then they retrieved their ancestors from the Smithsonian and placed the crates on reserved seats on an airliner. They rode next to them all the way home. Wagner said the delegation made an odd sight as it was ferried by sky caps' golf carts—crates, tribal elders and all—to make a connecting flight in Minneapolis.

At Great Falls, they loaded the bones into Weatherwax's half-ton Chevy pickup for the last leg home to Browning. The highway winds past Old Agency, where Danials had stolen these bones, at a site where warriors for generations have paused to celebrate success.

"We stopped and sang some other old songs; then last we sang some victory songs. We had a victory over a bureaucratic institution," Weatherwax said. "It's a cycle. We live in cycles. You go back to the earth that you came from."

This matter of returning to the wisdom of one's ancestors, of returning to the earth, is no small matter for Weatherwax. He is a Sun Dancer. Now each spring these plains again resound with the ceremony of the Sun Dance. There was a hiatus. White overseers of the reservation judged the ceremony barbaric and so banned its practice; the traditionalists have resurrected it. It is a gathering every year in June in a large encampment centered around a special, circular lodge, maybe fifty or so feet in diameter. There is much to this ceremony, much dancing, but its climax comes when some celebrants are suspended by leather thongs strung in slits in the muscles of their chests or shoulders.

As Weatherwax tells me about this, we are standing in the parking lot outside of the college. An early winter wind is punishing us, pushing bits of snow across the plains.

"I am a Sun Dancer. We have brought that back. We now consider our Sun Dance to be whole as it was given to us," he said. "Why did they nail Christ to the cross? They sacrificed him. He sacrificed the only thing he had, which is life, his skin. When you're sacrificing something, in order for it to be a really noble sacrifice, you have to give something you own. The only thing I have and own is my skin. Everything else, Mother Earth and the creator already own."

Beneath Weatherwax's Western shirt, scars record his conviction.

MY DISCUSSIONS WITH THE BLACKFEET HAD JUMPED BEYOND the bones to this now, to scars and matters of conviction and pride, to matters of integrity in one's own life. The traditionalists speak frequently of alcoholism, both as they have dealt with it in each of their lives and also as it weighs now as an epidemic in the younger people of the tribe. Alcohol, and increasingly drugs. The problem is everywhere on this and all reservations. The traditionalists have sweat out alcoholism, literally they say, in sweat lodge ceremonies, purification. They see this cleansing, however, not as the end, but a step on a much longer road. The sweat is an opening of the pores, ears, and eyes, a purging of the poisons that block communication with the earth. It is the touch of the earth, then, that ultimately becomes the antidote to the alcoholism and its several attendant plagues.

Just west of Browning the mountains rise, drawing that line I seem to keep encountering. The mountains here are known as Glacier National Park. To the south of the park they become the Badger–Two Medicine, part of the Lewis and Clark National Forest, federal land ceded by the Blackfeet. Two medicine, double medicine, a place of magic and power. To the Blackfeet, these are the mountains of creation. Their religion says creation—not just theirs but that of all people and all things—took place right here almost within sight of Browning. Their Eden is a short hike up a creek. Yet they hold creation to be a continuing process. These mountains still are the creator, still the mystery, a place where one periodically visits to be re-created. When senses have been purged, when an individual is ready to see, then he goes to these mountains for his vision. He sits in meditation and fasting for as long as it takes. If he listens, not with his preconceptions but with a mind tuned to the earth, he receives a vision that serves as the guiding force of his life.

Besides guiding spirits, the Badger–Two Medicine is said to contain oil and gas. The Forest Service's geologists' most optimistic estimates say the area holds enough hydrocarbons to satisfy less than a day's worth of the nation's appetite. They say there is about a one percent chance of actually finding even this much oil or gas. Nonetheless, the Forest Service, as of this writing, has proposed to allow Chevron and Fina Oil and Chemical Company to drill on the lands the tribe holds sacred. The oil companies hold valid leases, and in our world that is sufficient to allow them to attempt to drill, no matter what the odds for success. Some members of the tribe, at times even a majority of them, back the oil companies' proposal, seeing it as an economic boost in a poverty-stricken town. The traditionalist faction is usually a minority but is adamantly opposed to drilling or any form of development of the Badger–Two Medicine. They say there can be no spiritual guidance in a land that has been roaded, rigged, piped, and drilled. They claim they can hear nature speak clearly only on unspoiled land, as the Badger now is. It is outside of the nearby Bob Marshall and Great Bear wilderness areas and so without formal wilderness protection. Still, it is roadless and is being considered for formal wilderness designation. Unless the oil companies prevail, it will remain a quiet land where the spirits can sing.

When the traditionalists talk about this new battle, another common element of their individual histories emerges: they speak of being veterans or warriors. Now I bust in on a meeting Wagner told me about. A small group of traditionalists is discussing the oil-drilling issue with the Forest Service. The meeting is in progress when I arrive. The Forest Service people are instantly uncomfortable that I am there, especially when my notepad emerges. They discuss my leaving. I suggest a conversation with higher offices in Missoula. They decide I can stay. Then the discussion proceeds, not so much to settle the issue as to illuminate the existence of a great cross-cultural gulf. The Forest Service is trying to come to grips with the notion of sacred. Perhaps, they suggest, the Blackfeet ought to point to a specific place in the Badger–Two Medicine that grants it special status, maybe a specific rock, tree, or hillock, and then the Forest Service could protect that. The notion of a sacred stretch of land does not register.

The government people face the task of balancing rights that cannot be weighed against each other. The oil companies have a lease, a property right, a right our culture holds to be inviolable, even, some would say, sacred. How do we weigh that right against a culture's right to preserve a place to touch its history? How do we balance our addictive appetite for oil against the tribe's attempt to treat alcoholism with spirit? This attempt to weigh a property right against their religion angers the traditionalists. They see it as no contest. Their spokesman in this meeting is John Murray, a Vietnam veteran, a philosophy major at Montana State University, and keeper of a sacred bundle, a sort of mantle of spiritual responsibility. He is becoming increasingly frustrated with trying to cross this cultural gulf, and finally the frustration vents.

"This land is the last stronghold for our way of life. That's why it has to remain undisturbed. . . . You guys will have to kill me [to drill for oil]. That's the bottom line," he says. "We are native to this land. Our religion is native to this land. We are primitive. . . . Primitive and primary are one in the same."

I leave the meeting, convinced I have a story. A headline forms: "Pikuni traditionalists vow violence to thwart oil rigs." This is news. Still, I need another interview, one promised from the outset. Curly Bear had said I would get to speak with Buster Yellow Kidney, a pivotal figure in all of this. Buster is a generation older than the others. His war was in Korea. He is regarded as a leader by the younger group. In the hand-to-hand passing of the delicate thread of Blackfeet culture, Buster serves as a tie to an even older generation, people like Mike Swims Under. Buster kept the coals alive during the fifties and sixties to spark the resurgence of traditionalism in later decades.

Given his status, I speculate that I can get him to confirm and elaborate on this threat the younger men made. After two days of trying, I finally find Buster. He meets me in a conference room at the offices of the tribal housing authority, where he works. I ask him as quickly as I think prudent about the threat of violence. He shies from the question not at all, answers it as directly as if he were stating his name and address for the record.

"We have a plan," he says, going on to speak of his combat training in Korea, Murray's experience as a Marine, the former Green Berets among the traditionalists, their familiarity with and access to weapons. Yes, they would defend the Badger if it came to that, and everything in the tribe's history suggests it might. But now the questions begin to roll on to other topics. As Buster answers stoically, I begin to suspect I am missing something important. More and more, I am moved by the calm of this man. Buster is large and as imposing as Wagner. The effect is telescoped by an oversized head. His looming, grooved face earns its strength especially from a pair of dark, sad eyes. He projects at the same time power and peace.

I was immediately comfortable with him, but just as immediately convinced of a missing element, as if my questions were glancing off, how far off I am only beginning to understand. I was playing a reporter of politics by focusing on the strategy, the battles. To me, the questions at issue, both in the bones and in the Badger, were questions of agendas, of cause and effect. What steps have come until now, what steps next? What is your agenda? I am of a culture that values its present only as a predictor of the future. Our people are perpetually plotting and drawing a line that leads elsewhere. To the Blackfeet traditionalists this makes no sense. They are not so much concerned about plotting a future as they are with repairing the present. It seems as if there are holes in their hearts, places left vacant by an unfulfilled past. This is less about drawing straight lines to tomorrow than it is about completing the circles that wind the past to the present. These are circles of ancestors, in the case of the bones. In the case of the Badger–Two Medicine, these are circles that bind a people to place. These are, in the end, the same circles.

Buster was attempting to make all of this clear to me in answer to a profoundly stupid question of mine. Why do those bones in that church matter to you? I asked. He told me a story from his childhood, how he would visit a cabin that housed, as was the tradition at the time, the bodies of a family that had died. This is the story:

> They had moved a whole family—I don't know how they died—there was a man and a woman and two children. And they must have moved the whole cabin up on top of this big ridge, and they put them inside. Everything was just the way they left it, the stoves, the plates, the woman's frying pans. Everything was just as if they went to sleep. The objects were on the wall, the beaded stuff was on the wall, and they were dressed in their full-beaded outfits. And it was the high point of all the surrounding area, so early in the morning I used to go after all the horses. That was my job as a small boy, and it was about a mile and a half or two miles away, so I'd get up on the high point . . . and I'd look over, and I could see the horses from there.
>
> But I'd always go into the cabin. And it had a door. Everything was perfect in it. And I'd go in and sit down and talk to them as if they were alive. I mean, they were alive to me, and any time I was sad and some things didn't go my way, which is quite often when you're a small boy, . . . I'd be up there. I'd be up there sitting there crying to them and asking them to help me and stuff like that. I know a lot of times it would be just a feeling that would come over me, but I would forget this [problem] and I would leave.
>
> So they were very much a part of me in my early childhood. Then I started to go to school, and I had to walk so far to school and so my grandfather asked my aunt if I could live up there so I wouldn't have so far to go to school. So I lived with them. . . . I was pretty much up there all the

school year. Then in the spring I would go home, and as soon as I got home I would run up there [to the cabin], and I'd sit there for maybe a couple of hours, and I always brought little offerings of tobacco or something. Then one of these years I came back and that [cabin] was the first thing I looked up and it was gone. I mean the little cabin was gone.

And I got home and I asked my grandfather what happened. He said, "Well last winter, last fall before the snow, there were car lights one night. There were car lights." He says: "I went up the next day and all the skeletons were gone, everything in the cabin was gone. A few nights later there were car lights up there, and I saw a big fire. Somebody burned the cabin down."

So I went up there. It was really sad, as if I had lost people, live people, as if I had lost them. I went up there, and I felt so sad that I sat there for half a day crying because I missed them, and during this time I asked that in the future, let me be a part of bringing you back. Make it possible that I be the one to bring you home.

A violent theft of the past has left the Blackfeet adrift. I see them as if in a child's drawing, suspended, people whose legs do not reach all the way to the earth. In this, they are not that different from the rest of us, only their earth is so close. Eden is just a short hike away. The immediate presence of the earth makes the effect of this suspension, this gap between their land and their feet, more apparent. Their route to regaining a touch of the earth seems so short; ours seems so long. I can only imagine that this makes the wind wheeze harder through the holes in their hearts. A few miles south of where Buster and I sat was the piece of land where his people were defeated, killed and dismembered. Many Blackfeet pass the spot every day of their lives. A few miles west rise the mountains where Buster Yellow Kidney believes his people and all people began. These same sacred mountains, he tells me, his face stoic, his great hands spreading before my face like a hawk's wings, are home to Napi, the creator.

It is two years now since that conversation occurred. Now I think about what Buster believes of the power of the mountains. A dozen years ago I had never seen mountains, but on a whim almost, I decided to change that. I drove straight through from my hometown in Michigan. I came to a spot I had picked on the map, nearly at random, a spot in the mountains within sight of Browning. Just after that trip and probably for the first time in my life, I made a decision my brain did not fight: My wife, my son, and I left the Midwest to live in mountains. Then, I believed that one range of mountains was as attractive as another, and so moved to Idaho, three hundred miles south of the Badger–Two Medicine. Still, events in my life kept conspiring to move me, ever, it seemed, closer to Browning. I landed in Missoula almost six years after the move to Idaho. Still, I was three hour's drive from Browning. In Missoula, however, my life uncovered some of its fundamental questions. Exploring them seemed to pull me east and north to those mountains.

Now I think of forests, mystery, and creation, of death, of fire that pushes east to the plains, of continents colliding, burying volcanoes that had before buried dinosaurs that came from eggs, of a bit of land that draws a line between the conflicts of our time, between wilderness and civilization, between oil and missiles in the ground and grizzlies above it. And now these forces seem to have brought my life to a line of its own, and I have crossed it. Once I would have considered Buster's notion about creation to be a curious ethnocentric artifact of his culture. Now when he says the creator lives in these mountains, I cannot disagree. Having crossed that line, I find I cannot go back.

CHAPTER EIGHT

ON A WET MORNING in August of 1989, I returned to the cover of the forest. I was on assignment, my last for the *Missoulian*. As dawn broke, I scrambled out across a clearcut that was a tangle of felled trees. The sawyers had been there, but the skidders had not. Trees lay like a jumble of pickup sticks, tangled everywhere. I crossed this battleground on bridges formed by horizontal trunks that were slippery, wet, and cold, as my jeans were from the morning rain. At the opposite edge of the cut, where the saws had last worked, trees still stood. In three of those trees environmental activists had rigged plywood platforms a good sixty feet up. There they promised a stand against the progress of the saws. I knew a couple of them, a quiet philosophy student named Gus, a gregarious rugby player, Jake Kreilick. The third, a woman identified then only as M. B., now known to me as Mary Beth Nearing, I had not met. She is hard-core Earth First, but something of an anomaly in a group known for its sometimes loony or destructive methods. She is a peace activist, and eschews the controversial tactic of tree-spiking, the driving of huge steel spikes into trees to thwart saws. She is a follower of Ghandi and will not participate in a demonstration unless everyone else signs a pledge of nonviolence.

The sawyers, by coincidence, had taken the day off. Federal agents carrying guns and wearing bullet-proof vests did come, though. There was a calm, even good-humored discussion between the feds and the protestors. Then a man arrived from the mill that had contracted for the Forest Service logs. He stayed at the opposite edge of the clearcut, away from the protesters. I approached him and identified myself. He said he knew who I was and would prefer not to talk to me. By then the timber industry was nearly unanimous in its lack of appreciation for my work. After some preliminaries (actually, we hollered at each other at first) he finally agreed to answer questions. He gave me a tour of the clearcut, and claimed his sawyers were doing a good job of forestry. For a bit, I was inclined to agree with him. A lot of trees were cut, but a lot remained, some twenty and thirty years old, a decent beginning. I pointed this out. Oh those, he said. Won't be there long. Subalpine fir. Trash trees really. Dozers will knock 'em all down when they do the skidding. Clear the junk out to plant decent trees.

Such is the art of forestry. A species fails to register on this man's scale of values, and I had nothing more to say to him. Nor had I any way to write the significance of his callousness in the newspaper story that would be my product for that day. I would instead write how protesters said this. Marshalls and loggers said that. On to the comics and the obit page. I left the clearcut to go about my business. So did everyone else, except the tree-sitters. They remained a couple more days, then climbed down, either satisfied their point was made or satisfied of the futility of making it.

My reporting of the story was hollow, but now I believe its greatest fault was in failing to account for the bear I saw. I had risen about four A.M. that day for the hour-and-a-half drive from Missoula to the tree-sit. The site was near the spot where Michael Gallacher and I had seen that first grizzly bear when we set off to write about trees a year and a half before. It was still dark. I cut north from Highway 200 to head into the Seeley-Swan Valley. About ten miles further on, my headlights picked up a great, black heap beside the road. I stopped, got out of my truck and poked at the dead bear spread on the shoulder. It was the biggest black bear I had ever seen, not there long, probably hit by a truck just that morning. For some curious reason, I felt the need to touch it. I picked up a limp paw and rolled it over, then rested my palm against the pad of its foot and laced my fingers in the dead bear's claws.

THE TWO-YEAR PERIOD OF MY LIFE I HAVE BEEN DESCRIBING can be charted on the undulations of Montana's mountains or by the peaks of my own anger. To trace the latter, I flash back now to the zenith, a time of rage when my editors balked at the publication of my series of stories on logging. Curious how that anger waned.

The episode was steeped in evidence of corruption. My bosses and the newspaper business in general had banished the sense of purpose that had initially attracted me. Instead, the spirit of journalism had become identical to that of cutting trees, selling shoes, and running the nation's savings and loans. This transformation of the business was in a sense convenient for me. It allowed me to wallow in anger, to do nothing but shriek, "I accuse." I wrongly believed, as most people do, that I was absolved of responsibility for the system. Yet the mountains sent waves to counter my rage. The forest advised that I too was the system, the community. As my health was dependent on its integrity, so was the system's health dependent on mine.

The root indicator of the moral bankruptcy in the newspaper business is the degree to which it has forsaken its subversive role. I came to the business because it was subversive, inherently. The asking of questions, as we have known since Socrates, is a profoundly subversive act. It is the foundation of speaking truth to power, and it lies at the heart of the human ability to steer behavior toward something decent. It is the willingness to infuse the system with the fire that ensures its growth and health. My newspaper and my bosses were not asking questions, and I screeched at them for this. In this anger, I was wrong. Asking questions was not their job. It was my job. I had no right to rail against their timidity unless I was willing to face the consequences of my own questions. The forest told me this because the forest is, at bottom, a web of subversive questions. It is harmony derived from upheaval.

In the year following my work on the logging series, there came to me as if sent a brace of stories that began framing themselves in larger terms: the fires, the dinosaurs, the festival of death and creation in Canyon Creek, the Blackfeet, bears, and stars. These began as usual assignments, but there emerged in each an inexorable element of the sublime that widened my universe of questions. There was, in each of these stories, a vein that seemed to twist back on the core issues of each of our lives, matters of heart. Encouraged by the quiet confidence of the forest, I began to write about these matters. Each of the events I describe became a newspaper story, and in this asking and writing, I found a measure of peace. During that pivotal year, my respect for my craft rose and my anger subsided. I believed there was something vital and important about this new role of mine, as a seeker of unsettling questions.

To me, these were important questions, important stories, but they fell outside the visible spectrum of the newspaper. My editors ignored the new direction some of my work had taken. I welcomed this. It meant I would be able to go on asking in obscurity, which in the corporate world often translates as freedom. And so I became, almost paradoxically, satisifed with my job, and at the same time, hugely dissatisfied with my employers. The latter angst seemed almost irrelevant, once I had assumed responsibility for the direction of my own work. It became and still is my work, not theirs.

While I explored this new ground, I continued to report along more traditional lines, covering especially the efforts of the timber industry in the state. My meat-and-potatoes work was aggressive, and the industry reacted. I was excoriated in letters to the editor, subscribers cancelled, timber-company executives flew into town to complain. During this period, John Mitchell of *Audubon* magazine interviewed Plum Creek's Bill Parson. Mitchell recounted that conversation in the magazine:

> [Parson] said, "The Press tries to tell the story, 'Gee, if you cut the trees it's bad for wildlife, all the dirt runs down the hill and gets in the crick and the fish all die.' That's what they say. Now that Dick Manning over at the *Missoulian*. He just puts out smoke. He has a hidden agenda. He buys his ink by the barrel. He's . . . you know . . . he's a known environmentalist."
>
> Well, look, I said. You must see the *Missoulian's* publisher from time to time, one businessman to another. What does he say when you complain about Manning and his barrel of ink?
>
> "We have had some discussions," Bill Parson said. "And I'll leave it at that."

Although *Missoulian* managers have since denied feeling (much less caving in to) any timber industry pressure, I believe they did. I had always considered such sniping as Parson's to be the wages of aggressive reporting. I had been down that road before—walking with my editors. This was the new era of journalism, however, and newspapers, particularly mine, were cultivating a fear of fire. Others around the country were discovering, as Doug Underwood had pointed out in his important article in the *Columbia Journalism Review,* that corporate journalism had no room for troublemakers, especially if that trouble accrues on the desks of fellow

MBAs. I was and am a troublemaker, both within and out of the office. One cannot ask questions, as reporters are trained and bound to do, without causing trouble. And one cannot ask questions without sooner or later tasting hemlock.

The first signs of my paper's impending capitulation came in a memo to me from editor Brad Hurd. He accused me of "crusading" in the timber series and suggested I cease. At the time, I considered this only fallout from the battle about the series, the sort of scrape that also goes with the territory. That memo, which came during the winter of 1988, was the last overt sign of trouble. During the following summer, however, covert signs began to abound. It was the sort of stuff as visible as a billboard for anyone who has worked nearly fifteen years in a newsroom. Editors picked apart my copy. Straightforward stories, day-to-day efforts, got weird play. Suggestions drew strange responses, no responses. Editors allowed me to spend several weeks researching a planned series of stories about old growth, but Hurd wouldn't schedule publication. I assumed he was stalling, and I correctly assumed he had a reason.

I had worked with Hurd by then for more than four years. That included a temporary assignment as city editor, when I reported directly to him. He is about my age and a likeable enough fellow. Rumor has it he was something of a hippie in his younger years. Outside the office, he and I get along fine, even agree on most issues. He has a quirky sense of humor I appreciate. Inside the office, however, we were perpetually at odds on the matter of style. I am an adversarial reporter, believing that truth often emerges from conflict. Hurd hates conflict, both externally with readers and internally with the people in his charge. His style is avoidance, and he can delay a nasty decision for months. I had seen him do it before, and now he was showing all the symptoms again, only this time I was the obvious target. Finally I got word that there would be a meeting between Hurd and me, scheduled for two days after the Earth First tree sit. I took the intervening day off to hike to the top of a mountain with my ex-wife, and we talked about what I by then assumed would be a nasty confrontation.

Hurd was nervous in the meeting and so was I. It was obviously a full-dress personnel action because he had invited a witness. Hurd quickly came to the point of the session: he said I had done some decent work but my passions were beginning to show. Can't have passion. Would use my talents elsewhere (where, he never said), but he decided I would be pulled from the environmental beat. I could no longer write about nature. I could write, maybe, about politicians, still considered fair game in corporate journalism. I could not, however, be left in a position to offend the timber industry. Journalism for the sole sake of profit had arrived full force at the *Missoulian*.

Now I consider myself fortunate to have had that moment; few of us ever have the watersheds of our lives mapped so cleanly before us. It was just the right jolt at the right time. It pushed my life where it needed to go. That meeting with Hurd completed the circle that had been trying to draw itself around me.

Two arcs of the circle already had connected in the forest. That is, it occurred to me that there were few differences between corporate logging and corporate journalism. Both were exploitive and myopic, willing to sacrifice the long-term integrity of either the natural or the human community for short-term profits. Both were activated by the principle of distance; that absentee owners, lacking commitment to and knowledge of the intricacies of place, could order the exploitation of place. When the logs ran out and the trees wouldn't grow or when the political discourse of a community was desiccated by a vacuous journalism, there was always another place to exploit, the next place interchangeable with the last.

This notion of distance, though, freights an even more fundamental problem of corporations, a disconnection inherent and encoded in langauge. The word *corporation* derives from the Latin word for body, and that is just what they are, or more accurately, are not. They are an attempt at body, a synthetic body, a surrogate body for the people who hide behind the construct. Corporations are disembodied, calculating brains. It is this disembodiment that is the root trouble of our times.

Our bodies, to the extent that we listen to them, connect us to nature. Our bodies bear our senses, our sensitivity, the constant flow of messages that teaches and warns us of the damage we do. Our bodies are keenly connected to survival and care about the quality of what we eat, drink, read, and breathe. Our bodies feel pain and joy. Our bodies will die, and we know it. Corporations, the collective body, the false body, know none of these things. Because corporations have no body, no place, no death, they are disconnected from the notion of community, either natural or human.

There is a term for the sort of forestry that has plagued Montana recently—"timber mining." It is a strip mining of sorts, a peeling of a layer of trees, of life, without regard for the intergity of the natural community that supports trees. It is a taking without a giving, accepting life without accepting the life cycle and the creative power of death. A tree farm rises on false pretense: growth is a line to the mill, not a circle back to the earth. So a tree farm eventually founders, languishes, and dies, a house of cards. Life that seeks to sustain itself without death is inherently corrupt and doomed. Tree farming is a taking. What is required to sustain our relationship with the forest is a giving, a surrender to the order of the natural world. I am speaking not of conquest, but of a marriage. We depend on the forest to sustain us, but in this there must be an acknowledgement of its rules, that we will allow it to change us. We must enter the forest with our bodies, all of our bodies, allowing our vision to guide our hands.

The disconnection of our vision from our hands, the disconnection from our bodies, is corrupt. In this corruption, corporations, tree farms and journalism, are no different than our nation itself. Corporations, once a convenient abstraction, have become the whole of life. Instead of demanding that our corporations become more like us, more human, more embodied, we have become more like them. We have forgotten we are a part of nature. Now a once profoundly human endeavor such as journalism has become disembodied, unnatural, dispassionate, valueless, inhuman. Journalism ought to be concerned with the organic record of a human community tied to a specific place, ultimately a human community that is a subset of a natural community. Journalism ought to feed as well as feed on its community. Journalism ought to be particular to place. Newspapers ought to be founded on the notion that a relationship of integrity, a marriage to the human community, will produce a discourse that is the result of that relationship. Journalism owes its communities more than cookie-cutter copies of *USA Today.* Corporate loggers mine trees, really, mine humus. Corporate journalism mines humans. This is our epidemic. Because we no longer enter the forest with our bodies, we have forgotten that human, humus, and humility are related words, all related to a Latin word for earth. We have corrupted these words.

Yet this corruption is driven by an enticing sort of cowardice that infects each of us, at least in some way. We are each prey to a fear of the unknown. Managers are hired mainly to render the world predictable, to act on a situation so as to guarantee an outcome, a profit, from the future. It is that pressure for a predictable future that forces managers to ignore as intensely as they are able the possibility of mystery, the uncertainty that comes from surrender to a relationship. A manager does not establish job security with a sermon on the wonders of caprice. A manager's value is in his ability to control, at least in the short term. The hand that lies across the land rules in fear of losing control. Managers are not unique in this, but the rest of us call this control "security," and on the land and on people, we call it "taming." Trash trees out. Good trees in. A predictable and managable future laid out on an eight-by-eight grid.

All of this, although unspoken, was the root of my problem with Hurd. The spoken part of the meeting was a different matter. He droned through the bureaucratic niceties designed to cover a manager's ass in a litigious world. I stared from his glass-walled office out across the open newsroom: rows of desks, eight-by-eight, what must have appeared to Hurd as a world in need of taming.

JUST A FEW MONTHS BEFORE MY END AT THE *MISSOULIAN*, Edward Abbey had died. At the time, I felt compelled to mount a pilgrimage in his honor, although I'm not sure why. I had never been a doctrinaire admirer of Abbey. I respected some of his work, especially *Desert Solitaire,* and his irascibility and iconoclasm, but he was a difficult fellow to admire. Still, I did. When I found there was to be a memorial in the desert overlooking Arches National Monument in Utah, I decided to make the 750-mile drive. If nothing else, a road trip in May has value: work the winter kinks out, see some country.

After a twelve-hour, straight-shot drive south, I pulled into Moab in early evening before the sunrise memorial service. I asked around, and got directions to the top, through a wired-shut gate, up an abandoned and disintegrating highway—Abbey would have liked that—to a rim of rock. It overlooked desert that defined three hundred sixty degrees of horizon. It was a slickrock plateau, board-flat slabs of sandstone, some house-sized, some as big as city blocks. It stretched out in a plane dotted just sparsely by yucca, a few juniper, a few pinyon pine. It was near dark when I finally unrolled my sleeping bag in the balm of that evening. Full moon. No wind. Full-bore quiet. Maybe a hundred bodies peopled the desert around me, but there was none of the usual camp noise. No fires or smoke. No motors. Just a few hushed conversations. Off on the edge of a brief breeze, a guitar. On a rock outcrop to the west, a woman sat in full lotus. The fading sun flushed the sandstone red. In this peace, I slept.

The service at dawn brought more pilgrims and more hush. A string quartet played. So did folksingers. A rancher spoke. So did writers I greatly admire: Barry Lopez, Terry Tempest Williams, Wendell Berry. Still, I had a hard time listening, distracted or even hypnotized by that great stretch of desert to the east. As the service ended I surrendered to the lure. I took a walk and got lost for a while among the water-carved canyons and the folds in rocks. Curiously, a word, just a single word, kept cropping up, a word I had not brought with me, had no reason to bring to a memorial for a dead man. The word was freedom, which, as I have come to understand it, is simply another word for wild, for wildness, for will, for wilderness.

In his essay "Walking," Thoreau wrote: "The West of which I speak is but another name for the Wild; and what I have been preparing to say is, that in Wildness is the preservation of the World. Every tree sends its fibres forth in search of the Wild."

That, also, is what I have been preparing to say.

Thoreau had it just right, but most of us understand only half of this. That had been my problem. We have headed west all this time only vaguely understanding that our west is but a metaphor for the wild. Our people do indeed value the wild, as long as it is the west, still elsewhere. We value it, some of us at a distance. A few of us to venture into it. Almost none of us, however, has courage enough to allow it to venture into us. We prefer our west somewhere across a line, a line just beyond where we stand. The west is held at bay, always just a few hills away, and we are always headed for it. But when we arrive, it is not the west anymore. The space we occupy we must tame.

We have cut a Faustian deal in this taming, a deal that included the taming of ourselves. This is the price of security. Yet as surely as I sat in slickrock, I can no longer accept the half of this deal. I cannot claim to revel in the wild hand ruling all around me and still refuse to allow it to rule me. The cost of security, the price of denying death, is the death of the wild in us all. The wild has a plan, which is the best plan, which is no plan at all. The wild is a tuning of the forces of life, a tuning to a patient harmony. If I will listen with my body then these cells of mine can hum to this same transcendent chord.

IN THE CORNER OF A GLASS-WALLED OFFICE, A NERVOUS AND timid man in a tie is offering me a new sort of job. I can still work for the *Missoulian*. Just because Hurd doesn't want me to write about nature anymore doesn't mean he can't use my talents, he says. This is a non sequitur to me now. Nature is everything. How can I accept not writing about it, which is to accept writing about nothing. I would as soon promise to breathe without breathing air. The meeting has gone on maybe twenty minutes now, and I have heard very little of what has been said. The bosses notice I have been silent and ask me to speak. I say I will have something to say presently and leave the meeting. I write them a memo that says this:

> [Earlier] I was overwhelmed with a feeling of freedom. Funny how wilderness and freedom seem to go together, but I guess that's why it interests me.
>
> Where I went wrong is to forget that freedom comes with a price. It always has. We too often forget that the sheer, liberating joy of being able to say "fuck you" to the sons-of-bitches of this world is often paid for by a fist to the face. But sometimes, like now, even that bit of freedom is worth a few missing teeth.
>
> I suppose at this moment, my response to both of you should be one of those fine, old Anglo-Saxon curses, but I find that after years of the pettiness of this place, I no longer have it in me.

Instead, I will simply wish you a peace of your own while I move along to learn and write about what interests me. At this stage in my life, that's my vision of freedom.

I quit. I will work until Sept. 15, if that is your wish.

It was not his wish. Hurd ordered checks cut for me that same day, and in an hour I was out the door.

IT IS A FEW MONTHS AFTER MY LEAVING. I AM WITHOUT A JOB for the first time since I was thirteen years old. Stone cold broke. In earlier years, resumes would have been in the mail. A new job, another paper and another town. I am convinced, though, that journalism will be just the same everywhere, and so a move is futile. I am convinced, too, that this business of taking on a new town whenever I use up the old one is no different than using up stands of timber, and so just now, I will stay here. Instead of moving, I stand in my back yard and cut wood.

Half my life is gone. I have had a continuous and—recent events notwithstanding—successful career. Now it is gone. I had a marriage; it is gone. My life seems to be divesting itself of those layers of relationships that were surrogates for living life itself. I seem to have been factored to my prime numbers. Now I must forget the surrogates and begin the living of my life. Because of this, I have chosen to cut wood with a four-foot crosscut saw. This seems to be where this process of re-assembling my life wants to begin. I will consider our use of nature, by which I mean killing. This reconstruction of myself needs to examine the decision to take life from the earth. This is where my questions were headed all this time.

I have seen in nature grand forces that know nothing of my life. They walk a scale of time on which my alloted span does not even register. These forces do not fear catastrophe; they thrive on it. I understand this and so do not suffer the grand illusion of current politics. My purpose here is not to save the earth because the earth needs no saving. The earth will save itself. It will accept the abuse we heap upon it until that abuse tips the balances evolution has so patiently tended. It is doing so all around us just now. Then the earth will react to establish a new balance as it always has, through catastrophe and extinction. A new order will arrive. We humans may or may not be a part of it. We are after all, a young, and by any fair standards, wholly unsuccessful experiment. We are the source of the globe's troubles, so why should it not purge us? Global warming. Ozone depletion. Even a plague as inelegant as starvation. All are relatively minor catastrophes when ranked on the scale nature already has known.

This is not about saving the earth. This is about saving us. I am genetically encoded, as are most species, to be singularly interested in the survival of our kind, deluded as that may be. But I am also interested in what kind we are, and so more than survival is at stake. This is about salvaging a measure of integrity to brace each individual life. Integrity is a word used to mean a sort of honesty, but also to mean whole, to be complete. I mean it both ways. We as a people are unwhole, a disease rooted in the incomplete understanding of the forces that create and re-create us. We have forgotten that in our daily breathing in and out, in our eating, in our reconstituting the molecules that were a few minutes ago nature, now us, we are simply re-forming nature to become our bodies, fleetingly our bodies, then back around again. The forest is not separate. We are only the most immediate incarnation of the forest. Reassembling our lives begins with touching the forces that make us.

So I cut wood with a crosscut saw. With my own hands, I make the sawdust fly. Then I split the wood. I swing a maul, and if I have read the grain honestly, the maul's bit will find the correct spot, a tree's tender spot, and the wood will split in an exuberant burst. Then the pieces will heat my house. All of this is at once an act of joy and of sadness. There is joy because it is work, a directness of purpose that rings clear back through my ancestry of saws, hammers, and axes. This is how we sustain our lives, and I value my life. But there is sadness because this cutting is an act of killing. I hold no illusions about this. It matters not a bit that the wood was dead when it left the forest, that it was slash left behind by loggers. I cannot lay this killing to the loggers. My appropriation of even dead wood precludes the possibility of future life by robbing the forest of bugs, fungus, and humus. This is killing, but my life needs to go on and I must heat my house, so I do it with my maul, ax and saw. I suspect this is the root of integrity.

I have friends who sincerely believe otherwise. They believe the initial step toward integrity is environmentalism, activism, the correcting not of lives but of politics. I think of a party I attended just after I left the *Missoulian*. One of my last stories at the paper was about an environmental battle being waged by a group of wealthy residents of a small lake in the Seeley-Swan Valley. Plum Creek had promised to install a fresh clearcut on its lakeshore land opposite the expensive homes, and so the country-club set joined the environmental movement. It made a fine story, largely because of the consciousness-raising the threat of the clearcut prompted. Scions of the Republican Party were huddling with some of the more radical environmentalists I knew.

The party I attended was a fundraiser for this battle and was staged at the home of fashion designer Liz Claiborne. She was there, mingling. So was her husband, Arthur Ortenburg, a man with an agile political mind. I enjoyed my conversation with him, but kept being distracted by the building we were in. It was an indoor riding arena the size of an aircraft hanger, easily the biggest barn I've ever seen. I was fascinated with how it was built, especially how it supported its massive free-span roof with trusses, wood trusses. Lumber. Trusses that appeared to be spaced along each foot of the building's span. This shelter of a place to ride horses probably held enough lumber to frame a dozen loggers' homes. I guessed it held nearly as much lumber as Plum Creek proposed to cut on the nearby hill.

There can be no questioning Liz Claiborne's and Arthur Ortenburg's commitment to the cause. The couple has spent millions of dollars on environmental projects. Still, that roof suggests that there is more to this matter, that somehow we must come to grips with use. This anecdote, after all, is only a matter of scale. I hate what logging has done to Montana, yet I use wood. My livelihood is words printed on paper. I use, and wish to make no case for denying it.

That denial is simply a counter-arrogance to that of developers. I have friends who claim they do not kill because they are vegetarians. I have friends who are just as arrogant as any timber beast in this regard. Can this be right? A wheatfield is a monoculture, an undoing of nature. The entire middle of our nation was once a vibrant and complex ecosystem, the short- and tall-grass prairies, every bit as exquisitely evolved as the forest. Now there is even less natural prairie left than there is old-growth forest. Now it is wheat. A wheat field is a Midwestern clearcut.

The American farmer spends ten calories of energy—hydrocarbons, gas and oil—to produce each calorie of crop. The "primitive" farmers of New Guinea reverse that ratio. A loaf of our bread is not made just of wheat, it is mostly made of fuel, and oil kills.

Vegetarianism is a bias toward animals, but nature knows no such bias. The web of life values equally the life of a fungus and the life of a deer. Nature does not spare a life only because it has big, brown eyes. I find no comfort in denial of my killing, only an avoidance of the issue at hand. I find no value in arrogance here, when it seems it is the opposite quality that the earth now demands of us.

It seems the issue is not whether we kill, but how we kill. At the very least, it seems we ought to kill with some respect for what we know, with some intelligent consideration of our long-term interests. This nation will continue to cut trees. There is no way we will or should stop, but in doing so we can begin to pay some respect to the mystery of the forest. The scientists like Jerry Franklin who pioneered some of the initial investigations into this mystery have also begun to design a system of forestry that uses what they have learned. Even Plum Creek officials have taken note of this, have begun to make noises, not about tree-planting but about attention to the integrity of the forest. Here are steps we ought to take, but take cautiously. We are babies in this matter. We know so little of what there is to know, so any system of new forestry is bound to be incomplete. We will make mistakes. No matter what we do, it will in some sense be wrong.

It is important that we honestly try to do what is right by the forest, but now it seems even more important that we also realize whatever we do is wrong. This is the case for humility that the forest makes best. The forest is a wonder beyond our comprehension. My friends who are biologists tell me there is a new theme emerging from their research. It is becoming clear that the more we know, the more we find we don't know. Our growing knowledge mostly illuminates our ignorance. This is the greatest service of our growing knowledge. The forest is a lesson as much about the state of our comprehension as it is about the complexity of the forest. I once heard a Blackfeet man pray. His most earnest plea was for forgiveness for his mistakes that would damage the earth. There was no subjunctive in this prayer. It was a prayer of a doubly bowed head in that it assumed that his life had done and would do great harm. There is a sadness implicit in this assumption, and yet, as I come to approach it again and again, I believe it to be the sadness that heals.

I cut my own wood so that I might take this sadness and twist it about in my hands, greet it each night as I stoke the stove and watch the flames dissolve the lives of trees. This thrusts me deeply into the process of the second-by-second trade of molecules that makes a life, my life. This is sad, but the alternative is far sadder in its emptiness. It is not necessary that I burn wood. There are other ways to heat a house. My house has a gas furnace, and natural gas is cheap. I could simply twist a dial and forget the whole business. I have a vague notion that gas would come flowing from a pipe leading to the north and well out of my sight.

Perhaps, even my demand for gas would help make the oil companies' case that it's necessary to drill in the sacred lands of the Blackfeet. Maybe not, but in the twisting of this dial on the wall I have removed myself from the process. I have lost the direct link between my hands and the killing of this Blackfeet land. Now it is out of my sight. I am human and willing to tolerate this hidden killing better.

Instead of natural gas from Blackfeet country, I could burn oil from elsewhere. It doesn't matter where, as long as it is elsewhere. Oil is convenient and would deal me a warm house each night. Then I could retire in comfort to listen to the evening news of American soldiers exercising the mechanics of their craft in the Mideast, in defense of oil.

My woodstove pollutes the air of my community, which, like many mountain towns, has a severe inversion problem. I understand this, because my body breathes the smoke. I do what I can to ameliorate this by owning the least-polluting sort of stove and by not burning when the air is the worst. But what people really mean when they say my stove pollutes is that it pollutes here where people live. Is it better to heat with coal-generated electricity and send the acid rain elsewhere to foul natural communities? All forms of energy consumption degrade nature. I prefer my pollution here, where I can see it, where my body can feel it and begin to accept some of the responsibility for the consequences of my burning.

I could stop burning wood and do as most Americans now have done, as our system of production has urged us to do: remove from my sight the consequences of my life. To do so would be to ignore the teacher I need. Because I understand the consequences of what I am doing to this tree, I am more reluctant to do it. I am more likely to learn about the forest so I can gather my wood in the least damaging way. I am more likely to buy a more efficient stove and insulation, to sit a bit cold in a sweater. These steps are all less convenient than simply cranking up the heat, but now I have seen the killing. I will reduce that killing to a minimum. At the same time, though, I will understand that the killing has not stopped. My life has a price.

This is the lesson my body wants to absorb by burning wood, but this is only my particular avenue, derived from the particular circumstances of my life and my place. The same lesson comes elsewhere and becomes general when we understand the importance of connecting our bodies to those natural forces that sustain them. The troubles of our time all seem rooted in a disconnection from those forces, distance from the raising of our food, the cutting of our wood, the building of our houses, even the singing of our own songs. Our bodies need to touch these things, not so much as the answer to our troubles but as a means to reconnect our senses to nature, to raise the possibility of answers. To re-establish this connection is to understand that the daily, simple decisions of each of our lives have consequences. It is the collective weight of those individual decisions that has so burdened the natural world. When we use, we kill, and when we use more, we kill more. Responsibility for minimizing this rests on every one of us.

To remove nature is to make the killing come easier, so there can be more of it. A man can buy a package of hamburger in a supermarket and leave no blood on his hands. Then he can donate money to the humane society and believe he is kind to animals. An executive of Champion International can sit in his office in Stamford, Connecticut, glance at his inventory and cash-flow sheets, then order that inventory liquidated, never once considering it as a forest. A logger living, hunting, and fishing in Missoula, still equally dependent on the forest for his livelihood and his future, is less likely to make the same decision.

We have removed ourselves from the natural world, and so denied its reality. Now nature is an abstraction, easily reducible to paper, easily held off at arm's length, at continent's length, easily crammed into the shorthand of business that considers a forest inventory. The organizers of our lives are abstractions that forget a living forest is beauty, peace, and a pattern for the conduct of our lives.

We have ventured now so far into the unreal that we have abstractions for our abstractions. Money was once considered the way to reduce all commodities to their common denominator. It is the abstraction that took that first step away from contact with the natural world. But now money itself has been reduced to a blip on a computer screen, a symbol shielded by a symbol, a commodity that can be manipulated, inflated, and traded without any reference whatever to the needs it is supposed to fill.

The savings-and-loan crisis that swept the nation at the height of our profligacy was symbolic of the time. It was the pinnacle of our disconnection from reality. Money was created by manipulating money, rolling over loans without regard to the demand that created them or the consequences of their ever coming due. The crisis worked the margin, not so much the margin of profit as the margin of reality.

Loans were made to build houses that were not needed because there was money to be made in simply issuing loans. Then it all collapsed and there were images on our television screens of mostly built homes being bulldozed, shiny new studs snapping and crumbling before the blades. Those hallucinatory loans created a demand for lumber, and so trees died in Montana. Their bodies were carried away. Then the trees were studs on a train, and they filled the fresh demand. Then the house of cards fell, and along with it fell the real houses. Splinters of studs, once trees, now rest beneath a parking lot in Houston.

When certain native people hunted deer, they carried corn meal. They would offer this meal to a freshly killed deer by way of apology. Most tribes sang songs to the departing deer's spirit. Some tribes brought the whole deer home, into the house, and treated it as an honored guest. This ceremony acknowledged the cost of the celebrant's life. Killing a deer is a hell of a sad thing. I understand this better as I become older and approach my own death. I see it spread wider each year across the eyes of the deer I kill. Yet with my life now emptied and stunned as it is, I wish to face this grief. I will cut my own wood. I will shoot my own deer. I will dig my own carrots. All will be dead, but will be real.

There is a contradiction that forms the rift of the human condition, a seemingly cynical joke worked by nature in the evolution of our brains. In its stubbornest form the paradox is this: our survival is dependent on our doing no damage to nature, yet to survive each day we must damage nature. I feel often as if all of our human contrivances, all of industrialization even, is nothing but one great Goldbergian device designed to obscure this contradiction. We have alienated our lives from the forces that shape them in an attempt to deny our killing— ultimately, to deny our own deaths. Yet this alienation has done terrible damage. Denying our deaths carries with it the denial that we are nature, every single cell of us.

A reassertion of our bodies, our senses, is necessary now. We must look at the forest and see the damage our hands have done. If there is freedom and integrity for us, it lies in overcoming alienation, and so I saw wood. This is the sadness. Yet as I begin to live with this sadness, I have an odd sense that it is finite. It is but a small part of larger circle of joy. There is grief in the forest, but just ahead at ridgetop, a light is gathering in the dead trees.

BIBLIOGRAPHY

THE FOLLOWING BOOKS and articles either provided me with direct information or steered the writing of this book:

Berry, Wendell. *Standing By Words*. San Francisco: North Point Press, 1983.

Berry, Wendell. *What Are People For?*. San Francisco: North Point Press, 1990.

Bolle, Arnold. "The Bitterroot Revisited: A University Review of the Forest Service," *Public Land Law Review*. Vol. 10, 1989.

Busterna, John C. "Trends in Daily Newspaper Ownership," *Journalism Quarterly*. Winter 1988.

Caufield, Catherine. "The Ancient Forest," *The New Yorker*. May 14, 1990.

Cox, Thomas R.; Maxwell, Robert S.; Thomas, Drennon and Malone, Joseph J. *This Well-Wooded Land*. Lincoln, Nebraska: University of Nebraska Press, 1985.

Kwitny, Jonathan. "The High Cost of High Profits," *Washington Journalism Review*. June 1990.

Maser, Chris. *Forest Primeval*. San Francisco: Sierra Club Books. 1989.

Maser, Chris. "Life Cycles of the Ancient Forest," *Forest Watch*. March 1989.

Maser, Chris. *The Redesigned Forest*. San Pedro, California: R. & E. Miles, 1988.

Mitchell, John G. "War In the Woods: Swan Song," *Audubon* November, 1984.

Mitchell, John G. "Tree-sitters Protest as Old Redwoods Fall to Corporate Raider," *Audubon*. September 1988.

Stern, Eddie. "Crusader vs. Raider in St. Pete," *Columbia Journalism Review*. May/June 1990.

Thirgood, J.V. *Man and the Mediterranean Forest: A History of Resource Depletion*. London and New York: Academic Press, 1981.

Thomas, Jack Ward; Ruggiero, Leonard F., Mannan, William R.; Schoen, John W.; and Lancia, Richard A. "Management and Conservation of Old-Growth Forests in the United States" *Wildland Society Bulletin* 16(3). 1988.

Toole, K. Ross; and Butcher, Edward. "Timber Depredation on the Montana Public Domain, 1885-1918," *Journal of the West,* Vol. VII, No. 3. July 1968.

Turner, Melissa. "The Feud That Rocked The St. Pete Times," *Washington Journalism Review*. September 1990.

Underwood, Doug. "When MBAs rule the newsroom," *Columbia Journalism Review*. March/April 1988.

Warren, Debra. *Production, Prices, Employment and Trade in Northwest Forest Industries.* United States Department of Agriculture, Pacific Northwest Research Station.

Yanishevsky, Rosiland M. "The Rise of Plans and the Fall of the Old Growth," *Forest Watch.* June 1987.